100% NEW
TECHNOLOGIES
AND SYSTEMS

100% NEW TECHNOLOGIES AND SYSTEMS

GARCIA FORTUNE'S TECHNOLOGY AND SYSTEM A
COMPLETE TECHNOLOGIC BOOK FOR THE NEAR
FUTURE A NEW WAY OF LIFE

Fortune Garcia

To order additional copies of this book, contact:
Xlibris LLC
1-888-795-4274
www.Xlibris.com
Orders@Xlibris.com
138300

CONTENTS

CHAPTER 2

SPECIAL THANK YOU

I WOULD LIKE TO THANK EVERYBODY WHO PUSH ME TO PUBLISH ALL MY MANUSCRIPTS WITH NO STOPPING WORKING.AS LOT OF AS THEY ARE, THE WORKS WILL BE FOLLOWED.

A SPECIAL THANK YOU TO XLIBRIS MEMBERS WHO NEVER STOPPED TO GIVE THE BEST WORKS AS POSSIBLE TO HAVE EACH OF MY MANUSCRIPTS PUBLISHED

I WAS THURSTY TO HAVE MY BOOKS PUBLISHED AND XLIBRIS COMPANY GAVE THAT ONLY CHANCE TO HAVE ALL MY NOVELS PUBLISHED.

AS FAR AS I AM FROM THE COMPANY, IT'S LIKE I AM AS NEAR OF THE COMPANY AS POSSIBLE

fortunegarcia@yahoo.fr

GARCIA FORTUNE'S TEHNOLOGY AND SYSTEM

BEFOREWORDS

THIS BOOK IS HARDLY TYPED, DIRECTLY BY THE WRITER FORTUNE GARCIA,HIMSELF, IN HIGH SECRET LISTENING GOOD AMERICAN SONGS

MY TRUE WANTING,AFTER A LONG TIME THESE MANUSCRIPTS WERE STOCKED SOMEWHERE,I DECIDE MYSELF OR I MAKE UP ONE'S MIND TO PUBLISH THEM BEFORE OR AFTER THE NOVEL IN MEMORY OF SOMEONE THAT THE WORLD LOVE SO MUCH.

I PUBLISH "100% NEW TECHNOLOGIES AND SYSTEMS" AS FAST AS POSSIBLE TO HELP THE WORLD TO ADVANCE ON THE BEST WAY OF TECHNOLOGIES ASKING AND WANTING BY PEOPLE TRUELY FOR THEIR FUTUR.

TO IMPROVE THE FUTUR OF THE WORLD TECHNOLOGICALLY,I THINK THE WORLD IS READY TO RECEIVE THOSE NEW TECHNOLOGIES DISCOVERED BY THE NEEDS OF PEOPLE ALL AROUND ME.

HIGH RESEARCHES HAS BEEN MADE TO IMPROVE A NEW KIND OF TECHNOLOGIES AND SYSTEMS. IN ALL MY PUBLISHED NOVELS,THERE ARE A NEW TECHNOLOGY WROTE JUST TO HELP YOU MY WRITERS TO BEGIN TO UNDERSTAND MY TRUE WORKS ABOUT NEW TECHNOLOGIES AND SYSTEMS THAT I SPECIFY SINCE MY YOUTH.IT IS REAL,NOTHING HERE IS A COPY OR AN ADVANCED COPY.ALL HERE IS A BLACK AND WHITE WORKS

STRONGLY AND DIRECTLY DISCOVERED BY ME,FORTUNE GARCIA.IT IS THE REASON WHY THIS BOOK WILL BE SOLD AT A HARD PRICE JUST TO HELP YOU TO KNOW THE IMPORTANCE OF THE "100% NEW TECHNOLOGIES AND SYSTEMS "

EACH COPY WILL BE SOLD DIRECTLY ON THE WEB OR BY PAPER BACK OR LIKE BOOK AT : $ 99,99 US

SELLING FOLDERS IDEAS INSIDE THE BOOK BY NUMBER

fortunegarcia@yahoo.fr

GARCIA FORTUNE'S TECHNOLOGY AND SYSTEM

GARCIA FORTUNE'S TECHNOLOGY AND SYSTEM

fortunegarcia@yahoo.fr

VERY IMPORTANT:

IF YOU WANT TO BUY OR PURCHASE AN IDEA FOLDER INSIDE THE NOVEL" 100% TECHNOLOGIES AND SYSTEMS,FIRST YOU HAVE TO CONTACT ME,FORTUNE GARCIA BY

A) E-mail : fortunegarcia@yahoo.fr

B) TO BE ABLE TO MAKE A DIRECT TRANSFERT MONEY BY BANK TO BANK ON MY ACCOUNT NUMBER IN HAITI

I WILL SEND YOU THE ACCOUNT NUMBER BY E-mail

IT IS STRONGLY BETTER FOR ME TO BE CONTACTED BY E-mail

MY PHONE NUMBER IN HAITI IS STRONGLY DISPONIBLE

TEL : (509) 3406—0050.

PLEASE, YOU HAVE TO CALL ONLY FOR IMPORTANT CASE

C) AFTER THE COMPLETE PAYMENT,I WILL SEND YOU A CERTIFICATE OF BUYING OF THE TITLE AND NUMBER THAT YOU BOUGHT INSIDE THE NOVEL STRONGLY SIGNED BY ME, FORTUNE GARCIA .FOR THIS PART,I WILL ASK YOU TO SEND ME

YOUR ADDRESS AND PHONE NUMBER BY E-mail TO SEND YOU AN ORIGINAL BUYING COPY BY DHL

fortunegarcia@yahoo.fr

GARCIA FORTUNE'S TECHNOLOGY AND SYSTEM

- -

GARCIA FORTUNE'S TECHNOLOGY AND SYSTEM

fortunegarcia@yahoo.fr

ALL INSIDE THIS BOOK IS FORMALLY,STRICLY AND FORMALLY PROPERTY OF FORTUNE GARCIA AND COULDN'T BE ENTERED IN PROCESSING OF CREATION OR HIGH PRODUCTION FOR INDUSTRIES ONLY IF ONE THEM IS SELLING BY ME DIRECTLY AND DIRECTLY ONLY BY ME ,FORTUNE GARCIA ,NOBODY ELSE.

FIRST,I WOULD LIKE TO SAY TO THE PRINCIPAL STRANGERS INVESTORS AND TO WHOM WHO WOULD LIKE TO BUILD NEW INDUSTRIES THAT EACH FOLDER NUMBERED IS TO BE SOLD,AFTER READING THEM,IF ONE OF THEM INTERESTED YOU,JUST E-MAIL ME .

THIS BOOK IS SPECIALLY WROTE ALL THE WORLD TO KNOW NEW POWERFUL EXISTENCE OF CREATION OF TECHNOLOGIES DEVELOPING BY GARCIA FORTUNE IS ALREADY EXISTED READY TO BE PRODUCED.MY PHONE NUMBER IS

(509) 3406-0050

MY E-MAIL : fortunegarcia@yahoo.fr

THESE CREATIONS, NEW TECHNOLOGIES FOR THE FUTUR ARE WRITTEN BY ME DIRECTLY,GARCIA FORTUNE. TODAY,THE WORLD IS READY TO ENTER IN THE NEW SYSTEM AND TECHNOLOGIES THAT I WROTE A LONG TIME AGO.MY PRINCIPAL OBJECTIVE IS TO SELL SOME OF THEM AND WORK WITH THE SELLERS IF THEY WANT TO CREATE THEM FOR INDUSTRIES.100 % NEW TECHNOLOGIES WILL CHANGE ALL THE WORLD IF PEOPLE ALL THE WORLD USE THEM WELL FOR THEir DAYLY LIFE.NOTHING THAT I CREATED COULDN'T BE A DANGER FOR PEOPLE ALL THE WORLD AFTER THEY GET OUT FROM INDUSTRIES.IT IS THE REASON WHY EACH INVESTORS WHO WILL BUY A FOLDER OF THIS BOOK, IF IT IS FOR THE IMMEDIAT CONSTRUCTION WILL FIND ME TO SUPERVISE ALL THE FIRST PROTOTYPE

fortunegarcia@yahoo.fr

GARCIA FORTUNE'S TECHNOLOGY AND SYSTEM

LOVE IS THE ONLY ONE THING COULD MAKE YOU SUCCESS NOW.4-12-2011

LOVE IS SOMETHING YOU DON'T HAVE TO LOOK FOR

IT'S SOMETHING YOU HAVE TO BECOME

THE LORD IS MY SHIFFERD,I SHALL NOT WANT

GOD IS THE ONLY GIFT YOU CAN POSSESS

FOREVER AND EVER

fortunegarcia@yahoo.fr

23-03-2012 GARCIA FORTUNE'S TECHNOLOGY AND SYSTEM

MY DESIRE TODAY

IS DIFFERENT FROM THE FISRT TIME I BEGIN TO WORK IN HAITI

I WOULD LIKE TO DEVOTE MY ALL FUTURE TO WORK FOR STRANGER

IN DEVELOPING NEW TECHNOLOGY AND SYSTEM ALL AROUND THE WORLD

TO TRAVEL ALL AROUND THE WORLD WITH NO STOPPING TIME TO THINK ABOUT NOTHING

JUST ONLY TO PUBLISH ALL MY MANUSCRIPTS

IT IS A DESIRE COMPLETE INSIDE OF MYSELF HIGHLY AND TOTALY COMPLETED

I DON'T CARE OF WHAT WILL HAPPEN TOMORROW OR AFTER TOMORROW

THE ONLY THING WHICH INTERESTING ME ,IT IS ONLY MYSELF,MY WELL BEING

MY MANUSCRIPTS AND GOD

THIS IS WHAT MY HEART IS LIVING FOR

NOTHING ELSE

GOD BLESS MY SELFISHNESS TODAY AND FOR THE END

AMEN

1-07-2012

fortunegarcia@yahoo.fr

GARCIA FORTUNE'S TECHNOLOGY AND SYSTEM

whatever it is taken me ,i will go where i have to go

my life is a powerful work in the blood of jesus

it is what i am and no one could find it who i am in the miror

fortunegarcia@YAHOO.fr

GARCIA FORTUNE'S TECHNOLOGY AND SYSTEM

CHAPTER 1

LIST OF NEW CREATIONS, TECHNOLOGIES AND SYSTEMS DEVELOPED

13—MAGNETIC ONDE VIBRATION

14—LAZER PROTECTION

15—TECHNOLOGIC HOUSE

16—SCHOOL COMPUTERS

17—CLEVER HOUSE

fortunegarcia@yahoo.fr

GARCIA FORTUNE'S TECHNOLOGY AND SYSTEM

1—A PROTECTED CAR :
SPY PROCTECTION FOR CAR

A SPY PROTECTION FOR THE CAR COULD BE PLACED INSIDE OR OUTSIDE THE CAR OR ANY KIND OF CARS.

THIS SPY PROTECTION IS HIGHLY CONSTRUCTED JUST TO GIVE A COMPLETE SECURITY INSIDE THE CAR FOR EVERYTHING BROUGHT.

TODAY,WHEN YOU LET YOUR CAR OUT IN THE STREET ,IT IS WITH TWO HEARTS.IT WON'T BE LIKE THAT NO MORE.

fortunegarcia@yahoo.fr

GARCIA FORTUNE'S TECHNOLOGY AND SYSTEM

2—A NEW GENERATION OF CARS

TALKING ABOUT A CAR YOU USE WITH ALL THE ABILITY NECESSARY TO BE USED IN THE BEST TIME WHEN WE NEED,IT IS SOMETHING HIGH.

A) FOR THE FUTUR OUR CARS WILL BE ABLE TO FLY

B) IN THE SAME TIME,THE CAR WILL BE ABLE TO BE TRANSFORMED IN HALF TIME LIKE A BOAT.

C) WITH A SYSTEM HIGHLY AND TIGHLY CLOSED,THE CAR WILL BE ABLE TO BE A SOUS MARIN VISITING UNDER THE SEA AS FAR AS POSSIBLE.

fortunegarcia@yahoo.fr

GARCIA FORTUNE'S TECHNOLOGY AND SYSTEM

3—THE FLYING CAR PART 1

WE TALK ABOUT GENERATION OF FLYING CARS A LONG TIME AGO.NOW THERE IS NO SUPPOSITION ABOUT ,IT IS TRUFULY AND POWERFULY ABLE TO BE CONSTRUCTED. AROUD THE FLYING CARS I SAW IN THE WEB,IT IS A MIND PLAYING OR A MIND JOKER THAT THEY ARE PLAYING WITH.THE RESOLVATION IS SIMPLE SO SIMPLE THAT THE CREATION OF A PROTOTYPE OF A FLYING CAR COULD BE CONSTRUSTED IN A FRACION OF SECOND. THE SYSTEM THAT MUST BE USED TO CREATE THE FLYING CAR IS HIGHLY MIXED.THE CREATION OF A FLYING CAr WILL BE SO SIMPLE to help the car TO FLY EASILY WITH NO PROBLEM.

fortunegarcia@yahoo.fr

GARCIA FORTUNE'S TECHNOLOGY AND SYSTEM

4—THE ROBOTS INSIDE OUR SOCIETY

THERE ARE A LOT OF WORKS WHICH ARE STRONGLY NEEDED TO BE DONE BY SOMEBODY OF SOMTHING ELSE. LIKE WHEN WE HAVE

A) THE HOUSE CARE COULD BE DONE BY THE ROBOT. EVERYTHING CONCERNS THE INSIDE OF AND THE OUSIDE THE HOUSE COULD BE CARED BY THE ROBOT.

B) THE SECURITY OF THE HOUSE COULD BE CONTROLLED BY THE ROBOT HELPED BY A SYSTEM DIRECTLY CONNECTED TO A COMPUTER, PERSONAL COMPUTER INSIDE THE HOUSE.ANOTHER PERFOMED SYSTEM COULD BE USED WITH THE ROBOT TO HAVE THE BEST AND STRONGEST POWER TO CONTROL INSIDE AND THE OUTSIDE THE HOUSE.

C) WHEN YOU HAVE TO BRING SOMETHING HEAVY,THE ROBOT COULD DO IT IN INDUSTRIAL WORKS,IN THE AIRPORT,THE WARF,EVERYWHERE WHERE THE MAINTNANCE IS HARDLY NEEDED A DOUBLE HAND TO STRENGHEN AND ACTIVATE THE WORK TO BE DONE FASTLY.

D) IN THE HIGHTECH CONSTRUCTIONS OF BUILDINDS OF TODAY AND FOR THE FUTUR,ROBOTS WILL BE A HAND TO GO FASTLY AND A DOUBLE IMAGINATION TO CREATE DREAMS OF HUMAN PEOPLE.

WE CAN MAKE A DIFFERENCE,A GREAT DIFFERENCE WITH DIFFERENT KIND OF ROBOT LIKE

      HOUSE ROBOTS INCLUDES BABYSITTERS ROBOTS

      CONSTRUCTION ROBOTS

      INDUSTRIAL ROBOTS

      ARMY ROBOTS

ALL DANGEROUS WORKS COULD BE TRANSFERED TO THE ROBOTS AFTER PREPARING IT FOR AND TO REALIZE IT.

ROBOTS COULD WORK IN FRONT OF US IN THE FUTUR LIKE BEST PATNERS OF HARD AND DANGEROUS WORKS.

fortunegarcia@yahoo.fr

GARCIA FORTUNE'S TECHNOLOGY AND SYSTEM

5—TECHNOLOGIC CITY

A NEW KIND OF CITY IS CREATED WITH ALL BEST NEW TECHNOLOGIES AND SYSTEMS SOMEWHERE IN THE WORLD.CITIZENS ARE LIVING BETTER AND WITH NO DOUBT ABOUT THEIR LIFESTYLE

fortunegarcia@yahoo.fr

GARCIA FORTUNE'S TECHNOLOGY AND SYSTEM

6—INDUSTRIAL COUNTRIES

A NEW KIND OF INDUSTRIES IS BORN WITH CLEAN SYSTEM EJECTED IN THE AIR.WITH THAT NON-PONLUATED SYSTEM IS RETOOK TO CREATE OTHERS NEEDS FOR THE CITIZEN FOR A COUNTRY WELL GIVEN.

WITH THOSE NEW NON-POLUATED SYSTEM OF INDUSTRIES THE BAD EFFECT ON THE PLANET EARTH WILL STOPPED TO HELP THE BLUE PLANET TO BE REGENERATED SLOWLY AND AFTER FASTLY .

WHEN HUMAN PEOPLE WILL ARRIVE IN A COMPLETE SYSTEM OF UNDERSTANDING OF THE UNIVERSAL WORKS AND ACTION,WE, HUMAN PEOLPLE WILL BE THE MOST WISE AS POSSIBLE TO SAVE THE PLANET EARTH OR THE BLUE PLANET.THIS WAY, THE BLUE PLANET WILL ABLE TO RUN ITS WAY TO LIVE A LONG,A LONG, A LONG , A LONG TIME FOR THE SURVIVED GENERATION COMING AFTER US.IT IS WHAT WE HAVE TO WORK FOR THE FUTUR,THE BLUE PLANET WILL ALWAYS BE GLAD TO SEE ITS SONS FOR SEVERAL CENTURIES AND ANOTHER MILLIARD OF CENTURIES.THAT'S WHAT YOU READ NOW IS ONLY POSSIBLE IF WE GIVE A CHANCE TO THE BLUE PLANET BY CHANGING OUR WAYS OF ACTING INSIDE OF IT.SOMETIME WE FORGET ABOUT THAT THE PLANET EARTH IS OUR HOUSE,A BEAUTIFUL HOUSE THAT GOD GAVE US TO PROTECT AND TO TO LIVE IN FOR GENERATIONS AND GENERATIONS COMING AFTER US.IT IS WHY GOD CREATE THE PLANET EARTH TO BE FOREVER

BUT SINCE THE HUMAN PEOPLE ARE ENTERING IN THE DOOR OF THE PLANET EARTH, THE PLANET BEGINS TO GET DOWN IN A FALSE WAYS THAT US HUMAN PEOPLE HAVE BEEN CREATED.

THE POWER OF TO BE WISE OR TO BE MORE CLEAR,HUMAN PEOPLE ARE AWAY FROM THE REAL POWER OF THE UNIVERSAL TRADING ON ALL OTHERS PLANETS.HUMAN PEOPLE WERE NOT ENOUTH IN CONTACT WITH THE UNIVERS OR WE,HUMAN PEOPLE DIDN'T LET THE UNIVERS INFLUENCE STRONGLY ON US TO WALK ON ITS WAYS OF KEEPING THE BLUE PLANET TOTALLY IN AN GOOG HEALTH INFLUENCING LIKE OTHERS PLANET DOING THE BODY OF THE UNIVERS.

ALL THE BAD WAYS OF TO BE WISELY KEPT BY THE UNIVERS WHICH HAS A GREAT SPIRIT OF INFLUENCING,THE BEST INFLUENCING POSSIBLE FROM THE UNIVERS WILL MAKE US SEE THE WAY TO USE POWER OF THE PLANET EARTH A DIFFERENT WAY.

THE BLUE PLANET HAS SUFFERING,HURTING,DESTROYING INSIDE AND OUTSIDE BY OUR WAYS OF USING ITS POWERS.

BY THIS CASE THE PLANET EARTH,YES, THE PLANET EARTH IS A BROTHERHOOD OF A MILLIARD ANOTHER PLANETS.

BY THE SPIRIT BROTHERS AND THE DIRECT SELF CONTACTING OF THE PLANET EARTH,IT EXPLAINS AND LET THE OTHERS PLANETS FELLS ITS SUFFERING FROM US HUMAN PEOPLE.

BY THE SAME WAY, THE OTHERS PLANETS SUFFER WITH THE PLANET EARTH AND THAT SPIRIT OF UNIVESAL POWER THEY POSSESS LET THEM KNOW IF THE PLANET EARTH DON'T LET OR HELP HUMAN PEOPLE TO LET THEM INFLUENCING BY THE POSITIVE POWERS OF THE UNIVERS,THE WORST COULD BE HAPPENED IN CERTAIN NEAR CENTURIES. HUMAN PEOPLE KNOW THAT

THEY WORK FOR THE DESTROYING OF THE PLANET EARTH,PREPARING A PLANET IN THE UNIVERS TO BE RECEIVED LIKE FOR EXAMPLE THE PLANET MARS,BUT THERE IS SOMETHING THAT HUMAN PEOPLE DON'T KNOW AND WHEN HUMAN PEOPLE WILL KNOW IT,IT WILL BE A COMPLETE DEFEAT FOR THAT PREPARATION OF TRAVELLING TO THE PLANET MARS.

THE HUNGRY OF THE SPIRIT OF THE UNIVERS IS THE WORST UNIVERSAL POWER THAT ALREADY DONE ON A LOT OF DESICION ON SOME DESTROYING PLANET.THAT'S IT,THE UNIVERS IN USUALLY CONTACT WITH THE BLUE PLANET DON'T STOP TO CRY AND CRY INCESSELY TO STOP OUR BAD USING OF ITS POWERS WRONGLY.

REMEMBER THAT

REMEMBER THAT ONE DAY

REMEMBER THAT THE BLUE BLANET COULD BE REGENERATED IF WE,HUMAN PEOPLE STOP TO USE ITS POWERS WRONGLY.THE BEST IS TO STOP AT A BETTER TIME.

THE UNIVERS COULD BE REGENENATED THE BLUE PLANET

THE UNIVERS IS ABLE TO BE REGENERATED INFLUENCING ON BLUE PLANET

THE UNIVERS WILL REGENERATE THE PLANET EARTH,THE BLUE PLANET.

fortunegarcia@yahoo.fr

GARCIA FORTUNE'S TECHNOLOGY AND SYSTEM

7—CONTROL INTERNET

EVERY COUNTRY TRY NOW TO CONTROL INTERNET IN THEIR OWN PLACE TO ERADICATE VIOLENCE OF EVERY KIND.THERE ARE SOME INTERNET SITES WHICH ARE VERY DANGEROUS FOR SOME COUNTRIES AROUND THE WORLD. THESE SITES SHOW FOR EXAMPLE CREATION OF BOMB OF EVERY KINDS SURELY LOOKING AT BY YOUNG PEOPLE OR CHILDREN.NOW A CREATION OF A POINTER TO KEEP THEM UNUSED WHEN THEY ENTER ON THE SITE,THE POINTER OR A SPECIAL WORLD LOGICIAL ALWAYS CALL POINTER COULD CONTROL EVERYBODY WILL ENTER IN THOSE SITES AND WATCH

THE REASONS THEY ARE INTERING AND THE SAME TIME STOPPING THEM IN A FRACTION OF SECOND DURING THEY ARE CHATTING.THE VERY IMPORTANT IS TO ERADICATE VIOLENCES POWERS AND VIOLENCES ALL THE WAY ON INTERNET SITES.THE WORK OF THE POINTER WILL HAVE TO:

*   TO DETECT WHOSE ENTERING IN BAD SITES

*   TO FOLLOW THEIR MOVEMENT ON IT

*   TO SEE WHAT THEY DOWNLOAD

*   TO REACT FASTLY

* TO GIVE INTERNET A SPECIAL POLICE WHICH COLLECT SELF AND DIRECT INFORMATION ABOUT THE USER ABOUT EVERYTHING ABOUT THAT PERSON.

* TO INTERCEPT THE COMPUTER USING BY THAT PERSON AND ASK TO THE USER QUESTIONS AND GO ON HAVING EYES ON THAT PERSON

* TO GIVE TO THE POINTER ONE DAY, THE POWER TO DESTROY ALL REFUSES SITES AROUND INTERNET AND CALL THE POINTER TO ERADICATE ALL NEW BAD CREATED SITES BY ANYONE ALL AROUND THE WORLD

* TO CREATE AN INTERNATIONAL FEDERATION POLICE OF ALL COUNTRIES ACCEPTED TO BE CONTROLLED FROM INTERNET BAD SITES

INTERNET MUST NOT BE A POWER OF DESTRUCTION, BUT A POWER OF CREATION FOR ALL COUNTRIES ALL AROUND THE WORLD.

INTERNET IS A BIG, A GREAT WINDOW OPENED ON EACH COUNTRIES WHERE THE SAME TIME, OUR ADVANCED CHILDREN,YES,OUR STRENGHEN ADVANCED CHILDREN USE INTERNET TO STUDY AND TO KNOW A LOT OF BAD THINGS.OUR CHILDREN COME AND PREPAIR THEIR FUTUR. IF THEIR FUTUR WAS DISTUBED BY BAD THINGS COMING FROM INTERNET ,THEY COULD USE IT ONE DAY AGAINST US AND DESTROY OUR COUNTRIES,CHILDREN, OURSELVES,PEOPLE AROUND THEMSELVES DIRECTLY.

LET'S SAVE AND PROTECT OUR FUTUR

fortunegarcia@yahoo.fr

GARCIA FORTUNE'S TECHNOLOGY AND SYSTEM

8—LAZER HOUSE SECURITY

SO MANY TIME,WE USE FENCES TO CIRCLE OUR HOUSE LOOKING FOR SECURITY NVER COME AROUND AND PEOPLE ENTER WHEN THEY WANT IN THE PRIVATE HOUSE. TO CONTROL COMPLETELY THE SECURITY OF OUR HOUSE AND BY ONLY ONE KEY THAT NOBODY WON'T HAVE . TO REALIZE A TOTAL SECURITY AROUND A HOUSE, IT IS EASY AND WON'T TAKE ANY GREAT PLACES AROUND THE HOUSE.

A SIGNAL CLEARLY WRITTEN WILL BE ON THE INVISIBLE FENCE

WITH RED SIGNAL TOO

A COMPUTER VOICE WILL STOP ANYONE TRYING TO GO THROUGH THAT FENCE

THE POSITION MUST BE COMPLETELY CIRCLE WITH LAZER WALL COMPLETELY INVIVISIBLE ON YOUR OWN COMMAND TO KNOW HOW YOU WANT IT.

THAT LAZER WALL OR FENCE WILL BE ACTIVATED WHEN YOU WANT IT AND WHERE YOU WANT IT AT THE TIME YOU WANT IT.

THE LAZER DOOR OF THAT POSITION WILL BE OPENED BY ONLY ONE LITTLE BOX THAT SOMEONE POINT ON IT TO ENTER AND CLOSE. A RADIO SYSTEM COMPLETELY WITH

LAZER WALL WILL BE ADJUSTED TO PREVIEW EVERYDOBY
THAT CROSS THIS FENCE IS HIGHLY DANGEROUS,TO SPEND
THROUGH COULD KILL SOMEONE WHO TRY TO SPEND
IT.FOR CHILDREN,DIRECTLY FROM THE LAZER DOOR WALL

A FALSE OBJECT PLAYER WITH LAZER BEFORE AND AFTER
TO KEEP THEM FAR AWAY FROM THE DANGEROUS LAZER
WALL. THE LAZER BY BY GREEN COLORS WILL STAND
PEOPLE TO LOOK AT AND TOUCH THE LAZER WALL.

THE MORE IMPORTANT IS TO CONTROL LAZER SYSTEM
WILL HAVE A WRITING TO PREVIEW AGAIN PEOPLE TO KEEP
THEM AWAY WITH THEIR CHILDREN.

THAT KIND OF PROTECTION COULD BE USED BY POLICE
TO CONTROL A ZONE AND EVERYTHING NEED TO BE
PROTECTED

fortunegarcia@yahoo.fr

GARCIA FORTUNE'S TECHNOLOGY AND SYSTEM

9—CLEVER AND SPEAKING COMPUTER

EVERYBODY THINK ABOUT IT A LONG TIME TO CREATE AN INDEPENDENT COMPUTER,BUT DEPEND ON OURSELF CONTROL.FAR OF THAT REASON,THE COMPUTER MUST HAVE A BRAIN,AN ACTIVE BRAIN TO THINK TO ITS OWN USED.A MIND WITH CAPACITY TO THINK ABOUT AND CREATE IDEAS,ITS OWN IDEAS.THAT'S MEANT TO GIVE IT A GREAT ABILITY LIKE OUR HUMAN BRAIN,A GREAT DEAL OF WORDS.TO DO IT IN ITS MEMORY TO THINK,AFTER TO SPEAK BY ITSELF.TO SPEAK,THE COMPUTERS ALL AROUND THE WORLD ARE ALREADY ABLE TO TALK BY A DIRECT COMMAND OF READING THAT WE ASK TO THE COMPUTER,THE COMPUTER JUST NEEDS A RECEIVER FOR ITS VOICE OUT,THE SPEAKER FOR SONGS CAN BECOME IT.LIKE THAT OUR COMPUTER COULD BE CONTROLLED BY VOICE COMMAND FOR EVERYTRHING.ANYONE COULD TALK TO ITS COMPUTER WITH NO PROBLEM. EVERYBODY WON'T HAVE POSSIBLITY TO COMMAND BY VOICE ONLY,THESE ARE SOME PEOPLE IN TROUBLE CAN'T USE THEIR OWN VOICE TO USE THE PERSONAL COMPUTER,THAT MEANS,THE LAST INVENTION,CREATION STORY FOR THE COMPLETE COMPUTER.IT CAN BE SPECIALLY ENTERED WITH THAT NEW SYSTEM FOR HANDICAPED PEOPLE WANT TO HAVE IT WITHOUT ITS ALL MATERIALS,LAST CREATION MATERIAL LIKE THE MOUSE , THE TOUCHING POINT AT. EVERYBODY COULD BUY IT COMPLETELY OR WITH THE SYSTEM ONLY.

THE PERSONAL COMPUTER CAN SPEAK DIRECTLY BY SPEAKER AND HEAR,THE COMPUTER WILL BE ALWAYS ABLE TO CAPT ANY SIGNAL THE SAME TIME TOO BY THE SAME SPEAKER OR SOMETHING ELSE.TO MEMORIZE A SELF VOICE FOR ITSELF,THE COMPUTER COULD DEPEND ON THE PERSON BUYING IT.AROUND THAT, THE COMPUTER,THE NEW PERSONAL COULD BE SERVED OR USED TO CONTROL THE HOUSE COMPLETELY,CHILDREN,FOOD STORAGE,EVERYTHING TO MAINTAIN AND PROTECT THE HOUSE

THE EYES OF THE COMPUTER OR OF THAT NEW NEW VOICE OF THAT NEW PERSONAL COMPUTER WILL BE THE LITTLE VIDEO CAMERA ABLE TO ROTATE ITSELF EVERYWHERE IT NEEDS TO CONTROL,IT COULD SEE THROUGH WALLS AND EVERYTHING UNTRANSPARENT TO CONTROL TOTALLY THE HOUSE AND ALL AROUND IT OUTSIDE AND INSIDE.THE VOICE OF PERSONAL COMPUTER + THE SYSTEM OF WORKING DIRECTLY BY TOUCHING THE SCREEN WITH A SPECIAL BUNDLE OF STICK TO REALIZE THE COMMAND YOU NEED OR BY VOICE COMMAND THE NEW PERSONAL COMPUTER WILL BE ABLE TO COPY AN HANDWRITING SHEET WITH NO PROBLEM AND REALIZE ALL ELSE COMMANDS BY AUTO POWERS BY SELF VOICE COMMAND THAT MEANT A LIST OF WORKS COULD BE LET FOR THE WORKING COMPUTER

fortunegarcia@yahoo.fr

GARCIA FORTUNE'S TECHNOLOGY AND SYSTEM

10—SELF CONTACT BETWEEN THE NEW PERSONAL COMPUTERS WITH ROBOTS

ROBOT COULD BE THERE IN THE HOUSE TO ACT FOR EXAMPLE TO COOK,TO CLEAN THE HOUSE,TO WASH,TO SECURIZE,AT THE END TO DO ALL HOUSEHOLD NEEDED,TO HAVE MORE EYES IN THE CHILDREN TOO EVERYWHERE THEY ARE BY A PERSONAL PROTECTION PUT ON THEM,TO PROTECT THE HOUSE WITH COMPLETE WORKS OF THE NEW VOICE OF PERSONAL COMPUTER .IF THERE ARE FIRE,THE ROBOT WITH THE HELP OF THE COMPUTER WILL LOCALIZE WHERE THE FIRE IS TO STOP IT OR ERADICATE IT TO SAVE THE HOUSE.THESE KINDS OF PROCEDURE OR WORKS ARE POSSIBLE DIRECTLY BY THE ROBOT ON SUPERVISING THE COMPUTER COMMAND AS FAR AWAY IT COULD BE POSSIBLE,AND COULD BE USED TOO FOR EVERYTHING ELSE LIKE TO CONTROL A GREAT DEAL OF PERSONS IN A CONCERT WHERE IF SOMEONE LOST SOMEONE,ONE COuLD FIND EACH OTHER WITH NO PROBLEM

fortunegarcia@yahoo.fr

GARCIA FORTUNE'S TECHNOLOGY AND SYSTEM

11—THE PYRAMID HOUSE —THE FLYING PYRAMID

IT WILL BE THE NEW WAY TO TRAVEL AROUND THE WORLD AND BRING MORE PEOPLE.IT COULD BE USED TO GO IN THE SPACE.IT WILL BE PROPULSED BY 2 OR 6 TURBO PROPULSOR TO MAKE IT GET UP ON ITS BASE,IN EVERY SIDE,IF WE WANT TO ADJUST SOMETHNG TO GO FASTLY,IT WILL BE BETTER.WE CAN USE ANOTHER TYPE OR MATERIALS TO KEEP IT LIKE A PLANE ON THE AIR.THE INSIDE CAN BE BUILT WITH 1 OR 2 FLOORS.

LIKE HOUSE, THE PYRAMID HOUSE IS USED IN THE OLD TIME TO KEEP DEAD PEOPLE.BUT TODAY,WE COULD USE IT LIKE BUNKER OR LIKE HOUSE TO LIVE.IT WILL HELP US TO BE SAFE AGAINST EARTHQUAKES AND ANOTHER TROUBLES OF THE PLANET EARTH.

fortunegarcia@yahoo.fr

GARCIA FORTUNE'S TECHNOLOGY AND SYSTEM

12—THE CLEVER ROBOT SATELITE

EVERY DAY,WE SEND OUT OF THE WORLD SO MANY TECHNOLOGIES SATELLITES FOR INTERNET,RADIO,TV,CELLULARS COMPANIES FOR THE ONLY NEED TO COVER A PART OF THE WORLD.NOW WE THINK ABOUT A NEW KIND OF SATELLITES,NEW KIND OF CLEVER SATELLITES.IT CAN MOVE WITHOUT ANY HELPS FROM EARTH FROM ANYONE AND CAN COVER OVER ALL THE WORLD.BY ITS ROBOTIC POWER,IT IS ABLE ALSO TO COMMUNICATE DIRECTLY WITH US TO CONTROL ENTERING AND OUT MACHINES TO PROTECT THE WORLD,WAR DESASTERS FROM EXTRA—WORLD.

fortunegarcia@yahoo.fr

GARCIA FORTUNE'S TECHNOLOGY AND SYSTEM

A) COMMUNICATION

FIRST,THE SATELLITE HAS TO BE A ROBOT BY ADJUSTING A CAPACITY TO MOVE,TO TALK,TO SEE THE MORE TO BE ABLE TO REALIZE A GREAT DEAL OF WORKS WITHOUT HELPS OF HUMAN PEOPLE ALL THE TIME.THE SATELITE HAS A COMPLETE ROBOTIC POWER THAT LEADS IT TO A COMPLETE REALIZATION.COMMUNICATION WILL BE THE FIRST THING VERY IMPORTANT BETWEEN HUMAN PEOPLE AND THE SARELITE.THE SAME SATELITE COULD ENTER IN THE WORLD TO BRING SOMETHING NEW FOR US TOO,IT COULD TRAVEL ALL AROUND THE UNIVERS TO

DISCOBER,TO RESCUE,TO TAKE CONTACT WITH ANOTHER PLANETS WITHOUT ANY TROUBLES TO GIVE AND BRING US INFORMATIONS WE NEED DURING ITS TRAVEL DAY BY DAY, SECOND BY SECOND JUST KNOW WHAT IS GOING ON STEP BY STEP.EVERYTHING THE TRAVELLING SATELLITE WILL SEE AT THE SAME TIME NEAR FRACTION OF SECOND IS COMMUNICATED TO US.THE SAME SATELLITE CAN HELP SPACE BASEMENT WE SEND TO EXPERTIZE THE UNIVERS TO BRING AND SEND US INFORMATIONS THAT WE SEND FOR THE BASEMENT TOO.

B) ENERGY

THE SATELLITE WILL BE CONSTRUCTED LIKE A PYRAMID JUST TO HELP THEM TO BE LIGHT RECEIVED COMPLETELY FROM THE UNIVERSAL ENERGY POWER THAT HUMAN PEOPLE NEVER HAD KNOWN TO USE SINCE HIS COMPLETE CREATION.BUIT WITH A SPECIAL METAL,WE WILL STUDY SPECIALLY FOR ITS TRAVELS ALL AROUND THE UNIVERS,THE ENERGY WILL BE CAPTED ON THE UNIVERS LIKE A SYSTEM OPENED TO CAPT ALL ENERGY LIGHT ALL AROUND THE UNIVERS LIKE FOR EXAMPLE THE SUN ENERGY WHEN IT WILL BE AROUND THE WORLD,THE ABILITY OF THE SATELITE WILL BE ABLE TO CAPT ANY SOURCES OF LIGHT ENERGY IN LIFE INSIDE THE UNIVERS,IT WILL CAPT THE SUN ENERGY TO SURVIVE.WE CAN IMAGINE OTHER KIND OF ENERGY DIFFERENT FROM THE EARTH TO HELP IT TO SURVIVE WITHOUT CONTROL OF HUMAN PEOPLE TO GO AND TO DO EVERYTHING WE WILL NEED IT TO DO EVERYTIME AND EVERYWHERE AT THE MOMENT WE NEED IT.

C) THE SATELLITE

IT WILL COVER ALL THE WORLD BY ITS GREAT CAPACITY TO SEND INFORMATIONS,CONTROL INTERNET AND SEND ABILITY FOR TV,RADIO LIKE CIRCLE AROUND THE WORLD. THE SAME SATELITE COULD BE A GUN AROUND THE WORLD TO PROTECT A COUNTRY OR SOME COUNTRIES UNITED OR COULD BE TOO A GUN AGAINST EXTRA-WORLD

WOULD LIKE TO CONQUER THE WORLD.AT THE SAME TIME THIS LAZER GUN SATELITE COULD DEFEND ITSELF DURING A LONG TRAVEL AGAINST ASTEROIDS ON THE NEAR NANO SECOND AND AGAINST UNKNOWN OR UNDENTIFIED EXTRA-WORLD.

D) THE ROBOT MACHINE

IT WILL BE A ROBOT TOO TO BE CLEVER AND INDEPENDENT FROM US LIKE A LITTLE FREE BIRD TRAVELLING ALL THE WORLD AND THE UNIVERS TO BE ABLE TO MAKE RESEARCH ABOUT SOMETHING IT SEES A LITTLE CONFUSE AROUND THE WORLD AND THE UNIVERS. FOR THIS ,WE HAVE TO GIVE TO THE SATELLITE CAPACITY OF A COMPLETE CLEVER COMPUTER TO BE WITH ALL ITS MIND TO BE A GREAT MEMORY TO SEE, THINK,TO MOVE, TO GO, TO COMMUNICATE,TO RESEARCH,TO KEEP, TO SAVE ,TO RESCUE,TO SEND INFORMATIONS AT A NEAR NANO SECOND TO THE EARTH.THERE IS A GREAT ABILITY THAT WE REFUSED TO GIVE TO OUR MACHINE WHEN THEY ARE COMPLETELY CREATED,IT IS WHEN HAVE MEMORY TO FELL, TO THINK BY DEDUCTIONOF ITS SELF-MACHINE IT IS,CAPACITY TO RESOLVE,TO DO SOMETHING OR NOT,ONE THING AGAIN WILL BE ALLOWED ITSELF TO DEFEND ITSELF AGAINST SOMETHING UNKNOWN IT SEES WHICH ATTACKS IT TO DESTROY IT,IT CAN SEE THE ATTACK BEFORE THE ENNEMY'FIRES ARRIVE ON IT .ITS DEFENCE WILL BE COMPLETE AND DONE WITH A GREAT ABILITY OF LAZER GUN HIGHLY SPECIFIED TO DESTROY THE ENNEMY OR ENNEMIES WITH A GREAT POWER OF FIRE DIRECTION.

E) MIND SATELLITE

THE SATELLITE MIND WILL BE CONTROLLED BY AN EXTREM CLEVER COMPUTER DIRECTLY CONNECTED ON IT,INSIDE OF IT,MAKING PART DIRECTLY OF IT,TO BE FAST AWAY ABLE TO WAIT FOR EVERYTHING TO SEE .THE SATELLITE COULD COULD BE AN OPPORTUNITY TO KNOW A GREAT DEAL OF THINGS WE DON'T KNOW ON

THE UNIVERS.A CAPACITY WILL BE PLACED TO STOCK ALL ENERGY IT RECEIVES AND INFORMATIONS IT HAS TO VIEW ANY TROUBLES IN THE FUTUR IT COULD HAVE.IT WILL BE PERFECT,THE PERFECTION IN ALL.ALL INFORMATIONS IT WILL HAVE WILL HAVE A FAST AND STUDIOUS LOOK FOR INSIDE OF ITS MEMORY BEFORE IT WILL SEND IT BACK ON DIRECTLY TO A STATION ON EARTH OR WORLD SPACE STATION CONNECTED BETWEEN US IN THE EARTH TO CHANGE INFORMAtion IT COULD BE DIRECTLY FOR ONE STATION,SPATIAL STATION WITH GREAT CAPACITY TO RECEIVE HIGH AND FULL INFORMATIONS THE SAME TIME TO STUDY THEM FASTLY TO SEND THEM BY COMPUTER ALL AROUND THE WORLD IF IT IS NOT CONFIDENTIAL TO COMMUNICATE TO US.THE SATELITE MIND AND MEMORY WILL BE COMPLETELY OPENED TO BE ADAPTED TO LEARN CAPACITY TO COMMUNICATE INFORMATIONS WITH THE EXTRA-WORLD THAT MEANS IT CAN SPEAK AN UNIVERSAL LANGUAGE OR DOWNLOAD THE PERSONAL LANGUAGE OF THE EXTRA—WORLD IT NEVER KNEW TO TALK WITH OTHERS MOVING MACHINE IN THE UNIVERS WORLD LIKE TO ENTER IN THE WORLD DIRECTLY OR IN THE UNIVERS.

fortunegarcia@yahoo.fr

GARCIA FORTUNE'S TECHNOLOGY AND SYSTEM

13—MAGNETIC VIBRATIONS

HUMAN PEOPLE BECOME MORE AND MORE THE GREATEST FITHER WORLDWIDE OF THE UNIVERS.WE SEE FOR THE FUTUR, IT CAN HAPPEN WORLD WAR ALL AROUND THE BLUE PLANET, IT WILL BE GREATLY A DESTRUCTION FOR THE UNIVERS OR FOR THE BLUE PLANET.A WAR WHERE EACH OTHER SEND ATOMIC,CHIMIC,BACTERIOLOGIC,NUC LEAR AND NOW THERE IS A NEW ONE WHICH IS TRYING,IT IS A WEATHER POWERFUL MOVEMENT GUN.IT WILL NOT BE A GOOD THING AGAINST THE WORLD WHICH WILL BE DESAPPEARED WITH HUMAN PEOPLE COMPLETELY.A WAR AGAINST EACH OTHER IS NOT A WAR AGAINST THE HUMANITY,IT'S WAR AGAINST THE UNIVERS,THE CREATION OF THE WORLD,IT IS LIKE WE REFUSE OUR EXISTENCE TO BE HUMAN AND TO BE EXISTED,EACH WAR MEANT THAT WE DON'T ACCEPT TO LIVE IN THE WORLD,WE DON'T WANT THE WORLD TO LIVE IN IT,WE DON'T WANT A PERFECT HOME LIKE THAT,EACH WAR,WE, HUMAN PEOPLE IS DONE MEANT THAT WE WANT TO DESTROY OURSELVES TO SOMETHING UNEXISTED FOR THE ENTIRE NON EXISTENCE OF THE WORLD INSIDE THE UNIVERS.A WAR WILL DESTROY NOT ONLY THE WORLD,BUT THE UNIVERS,HUMAN PEOPLE AND THE SPIRIT OF THE PERFECTION OF HUMAN PEOPLE.

TO STOP THOSE KIND OF WAR,A VIBRATION PLACED ALL AROUND THE PLANET IS ABLE TO CAPT THE BOMB AND REJECT IT OUTSIDE THE PLANET AUTOMATICALLY JUST TO KEEP IT WHITHOUT ANY DESTRUCTION.THAT VIBRATION

WILL STOP TOO ANY GUN SHOT IN SOME YEAR JUST TO HELP HUMAN PEOPLE TO MAKE UP ONE'S MIND OF WHAT THEY REALLY WANT, THE DESTRUCTION OF THE PLANET OR ITS EVOLUTION. THAT'S THE REASON THAT THE SYSTEM WILL BE INSTALLED OR INTRODUCED ALL AROUND THE PLANET TO KEEP THE WORLD FAR AWAY FROM WAR AND THINK ABOUT OUR FUTUR.

FOR THIS REASON, YOU HAVE READ A COMPLETE FOLLOWING CHAPTER ABOUT MAGNETIC VIBRATIONS AROUND THE WORLD WROTE IN SECOND PART IN THE BOOK : "PRINCESS DIANA".

fortunegarcia@yahoo.fr

GARCIA FORTUNE'S TECHNOLOGY AND SYSTEM

14—LAZER PROTECTION

DURING THESE LAST YEARS,SO MANY TROUBLES HAPPENED IN THE WORLD.LIKE DESASTERS FROM NATURAL EFFECT OF OUR BAD UTILIZATION OF UIVERSAL POWER OF THE WORLD. TODAY,WE BEGAN TO SEE THE PRINCIPAL REACTION OF THE WORLD,SO MANY KINDS OF DESASTERS ARE WAITNG FOR HUMAN PEOPLE DURING THIS CENTURY AND CENTURY WILL COME AFTER WITH ANY GUN TO STOP THEM.WEATHER CHANGEMENT WILL BE SOMETHING PROPER IN OUR EYES DAY BY DAY WITH NO ABILITY TO DO NOTHING.WE HAVE TO FACE THEM AND ATTACK THEM BEFORE THEY ARRIVE AND DESTROY THE WORLD LIKE THE TSUNAMI DESTROYED ALL ASIA COASTAN KILL 2 HUNDRED THOUSAND PEOPLE IN A FRACTION OF SECOND.IT WAS LIKE ASIAN SAY IT'S A TSUNAMI EQUAL DESASTER OR NATURAL CATASTROPH.HOW CAN WE PREPAR THE WORLD TO SEND AWAY THOSE CATASTROPHS?

WITH NEW TECHNOLOGY AND SYSTEM,A PLAN TO CONTROL THOSE DESASTERS WITH GREAT INSTALLATION ALL AROUND THE WORLD MACHINES COULD BUILD LAZER WALL PROTECTION AGAINST FIRE,CYCLONS COMING IN A RESPECTIVE ZONE OF THEE WORLD.THESE TROUBLES COULD BE PREVENTED BY REACTIONS OF THESE SPECIALLY LAZER PROTECTION MACHINES AND THE SAME TIME BY A GREAT DEAL OF PEOPLE ALL AROUND THE WORLD WILL BE SAVED.

LET'S RETURN IN THE PAST

A TSUNAMI COMES IN ASIA,WITH A GREAT WIND AND THE DESASTER GROWS UP IN THE SEA STEP BY STEP IT ARRIVES IN THE COAST.IMMEDIATLY,ALL MACHINES PLACE ALL AROUND THE WORLD IN MOVEMENT IN SITUATED THE TROUBLE ZONE ,SEND LAZER RAYS BEFORE THE MOUNTAIN OF WATER AND DRAW A LAZER WALL BEFORE IT .AT THE SAME TIME ,THE MOUNTAIN OF WATER IS STOPPED AND AFTER A FEW MINUTES BEGINS TO BECOME TO BE PLACED AT ITS RIGHT PLACE AND LIFE GOES ON LIKE IT IS USED TO BE

WITH TECHNOLOGY AND SYSTEM,A CHANGEMENT OF OUR WAYS OF LIFE TO SAVE THE WORLD, HUMAN PEOPLE AND THE UNIVERS MAKES ONE WITH US.

fortunegarcia@yahoo.fr

GARCIA FORTUNE'S TECHNOLOGY AND SYSTEM

15—TECHNOLOGIC HOUSE

FOR THE FUTUR, A NEW KIND OF HOUSE IS DESIGNED TO CONTROL NATURAL

DESASTER.THE HOUSE WILL BE BUILD WITH DIFFERENT NEW MATERIALS VERY STRONG AND SIMPLE TO BE MOVED

A NATURAL CATASTROPH IS COMING IN A COUNTRY AND DESTROY IT. THE SAME HOUSE THE SAME TIME REACTS BY CLOSING ALL DOORS ,WINDOWS AND ALL THINGS

CAN OPEN TO KEEP AWAY WATER FROM THE HOUSE .THE HOUSE IS UNDER WATER AND EVERYTHING IS LIKE IT IS IN A REAL NATURE ,NORMAL NATURE,PEOPLE LIVE WITHOUT NO TROUBLE INSIDE OF IT.WHEN IN THE HOUSE , A LITTLE DOOR IS OPENED TO LET A LITTLE WATER CAN SPEND TO EXPLORE WHAT HAPPENS AFTER THE DESASTER.

THE HOUSE WILL BE BUILT TO BE ALIVE UNDER WATER WITH PEOPLE INSIDE OF IT .

fortunegarcia@yahoo.fr

GARCIA FORTUNE'S TECHNOLOGY AND SYSTEM

16—SCHOOL COMPUTERS

TO PROTECT OURSELF IN THE FUTUR,A NEW KIND OF SCHOOL WILL EXIST.THE HOME SCHOOL DIRECTLY WILL BE MANAGED BY A COMPUTER.THE HOME SCHOOL WILL BE COMPLETELY LIKE A NORMAL SCHOOL.THE MASTER OF ALL COURSE WILL BE THIS COMPUTER GIVING EXPLANATIONS, HOMEWORKS,SCHOOLWORKS AND WORKS WITH PEOPLE.THE SAME TIME,COURSES COULD BE GIVEN THE SAME TIME TO A THOUSAND PEOPLE LEARNING THE SAME THING AT THE SAME TIME.TEACHERS NOW CAN BECOME NEW TECHNOLOGIES COMPUTERS WORKS YEAR MANAGERS OF THESE TEACHERS COMPUTERS.THE COMPUTER WILL BE ABLE TO CORRECT THE CLASS WORKS FASTLY AND EXPLAIN EASILY TO PEOPLE SOMETHING WHAT THEY REALLY HAVE TO UNDERSTAND IN ALL SUBJECT. WHAT IS THE MORE IMPORTANT,THE COMPUTER AT HOME COULD BECOME A SELF PERSONAL TEACHER FOR SELF WORK.THESE KINDS OF SCHOOL COMPUTERS COULD BE AT HOME OR IN A GREAT SCHOOL COMPUTERS TO FAVORIZE THE MOST,THE LARGEST NUMBER OF PEOPLE TO BE READY FOR A FINAL COMPLETE KNOWLEDGE. THAT WILL DEPEND ON SOMEONE OR THE FAMILY OF THAT PEOPLE .THESE COMPUTERS SCHOOLS CAN COMMUNICATE THE SAME TIME TOO TO DEVELOP NEW TECHNOLOGY AND SYSTEM ABOUT INSTRUCTIONS TO GIVE TO THEIR STUDENTS OR DISCUSS ABOUT ANYTHING FROM A SUBJECT WILL NEED MORE INFORMATION THAT ONE COMPUTER DOESN'T POSSESS TO GIVE TO ITS PEOPLE. THE MORE IMPORTANT,SCHOOL COMPUTERS CAN HAVE

INFORMATION BETWEEN THEMSELVES AND RECEIVE INFORMATION TOO FROM DIRECTLY FROM SATELITES TO BE DISTRIBUATED BETWEEN THEMSELVES TO GIVE TO THEIR STUDENTS AND PERSONALS RESEARCHERS. IT CAN HAVE A CETRAL SCHOOL COMPUTERS TO TREAT INFORMATION AND SEND THEM ALL AROUND THE WORLD THAT WE CAN CALL DIRECT INSTRUCTION ON LINE ALL OVER THE WORLD WITH NEW TECHNOLOGIES AND SYSTEMS

fortunegarcia@yahoo.fr

GARCIA FORTUNE'S TECHNOLOGY AND SYSTEM

17—CLEVER HOUSE

IN THE FUTUR,I PUT MY CAKE ON THE OVEN,I CLOSE THE DOOR OR WE CAN SAY ANOTHER WAY,MY ROBOT HOUSE WILL PUT THE CAKE INSIDE THE STOVE AND RIGHT AWAY,THE PROGRAMMED STOVE ENTERS THE COOKING TIME ITSELF FOR THE CAKE.AFTER,THE STOVE SAYS THAT THE CAKE IS READY,ITS DOOR IS OPENED AND IT PUTS THE CAKE OUT WHERE A ROBOT COMES AND PUTS IT ON THE TABLE.

I AM IN FRONT OF MY TV,I SAY ,OPEN,IT OPENS ITSELF LONELY.THE ASK ME,ON WHAT CHANEL YOU WANT TO BE,I SAY 11 AND IT TURNS ITSELF ON IT.

THE NEW TECHNOLOGY IS MOST RAPID THEN WHAT YOU THINK

BY MY MIND WITH A LITTLE THING PLACED IN FRONT OF MY BEFOREHEAD,WITHOUT TALKING,I ASK EVERYTHING TO THE ROBOT AND THE CLEVER HOUSE, THEY RECEIVE ME 5\5 THAT I WANT IN THE HOUSE,I HAVE.

I ENTER HOME,I ARRIVE IN FRONT OF THE DOOR .THE CAMERA OF THE COMPUTERDOOR SAYS TO ME,GOOD AFTERNOON,MR AND IT OPENS THE DOOR FOR ME,I SAY GOOD AFTERNOON AND IT CLOSES THE DOOR.

MY FRIEND COMES HOME.HE IS GOING TO KNOCK THE DOOR AND AT THE NEAR SECOND,THE DOOR SAYS,DON'T

NEED TO KNOCK MISTER,WHO ARE YOU? HE SAYS TO THE COMPUTER CONTROLLING THE HOUSE,I AM MISTER J.THE COMPUTER LIST THE FRIENDS OF THE HOUSE AND HE SEES HIS NAME,AT THE SAME TIME,THE DOOR COMPUTER TALK TO SOMEONE INSIDE THE HOUSE AND ASK IF HE MUST OPEN ITSELF OR IF THAT SOMEONE WANTS TO RECEIVE THAT PERSON IN THE HOUSE,DEPENDIND ON THE ANSWER ,A ROBOT COMES AND OPENS OR THE SAME HOUSE COMPUTER SAYS TO MISTER J THAT NO ONE IS THERE IF HE WANTS TO LET A MESSAGE THAT THE PERSONAL COMPUTER WILL REGISTER IMMEDIATELY THAT MISTER J WILL SAY YES THAT HE HAS A MESSAGE TO LET FOR THE OWNER OF THE HOUSE,IF NO,THE COMPUTER WILL WISH TO MISTER J TO COME BY CALL .

EVERYTHING ALL AROUND THE WORLD COULD SPEAK AND ANSWER SOMEONE TO OBEY,GIVE INFORMATION OR SERVICE DEPEND ON WHAT THE PERSON NEEDS,IT WILL BE BETTER FOR US TO GO FASTLY AND SLOWLY IN OUR RYTHM.

EVERY CREATION SPEAK ONE TO ANOTHER AND ANOTHER TO HUMAN PEOPLE

WITH NEW TECHNOLOGY AND SYSTEM,WE CAN MAKE EVERYTHING SPEAK ALL AROUND THE WORLD AND ALL AROUND US AND IN THE HOUSE TOO FOR AN EASY USE AND CONTROL IN THE KITCHEN,DINING ROOM,BEDROOM,LIVING ROOM,WASHROOM,RESTROOM,GARDEN AND ALL THE AROUND OF THE HOUSE.THAT WILL HELP US TO HAVE ANOTHER THING TO DO AND TAKE CARE OF SOMETHING MORE IMPORTANT.ALREADY,A COMPUTER COULD WRITE COMPLETELY A BOOK.THE MORE HUMAN PEOPLE GIVE RESPONSABILITIES TO ALL THING ALL AROUND HIMSELF,THE MORE HE WILL HAVE MORE TIME TO TAKE CARE OF HIMSELF AND PEOPLE ALL AROUND HIMSELF ALL OVER THE WORLD.THIS IS A BEST THING TO SEE,EVERYTHING SPEAKS LIKE MY SWEATER TELLING ME

IT'S TIME TO TURN ITSELF IN WARM MOMENT IN THIS COLD TIME WE JUST ENTER NOW AND IMMEDIATELY THE SWEATER TURNS ITSELF IN WARM,EVERYTHING TALKS IN THE HOUSE .A BEST CONTROL OF THE CENTRAL COMPUTER HOUSE FOR RESPONSABILITIES OF EVERY HOUSEWARE IN WORK.EVERYTHING IS DONE BY ITSELF DIRECTLY.WE CAN SAY WE HAVE A HOUSE DEPEND ON ITSELF ON CONTROL OF THE CENTRAL HOUSE COMPUTER. WE,HUMAN PEOPLE WILL RELAX OURSELF DURING EVERYTHING IS WORKING BY ITSELF VERY,VERY,VERY WELL.THE NEW TECHNOLOGY HOUSE IS NOT TO KEEP YOU LAZY,BUT TO HELP YOU AFTER A GOOD WORK DAY WITH GREAT TIRENESS TO STAY QUIETLY QUIET AND RELAX YOURSELF AND YOUR FAMILY DURING YOUR MEAL IS PREPARED BY A ROBOT AND SET THE TABLE,A ROBOT BRINGS YOU YOUR HOT SMELLING COFFEE,AFTER YOU'RE TAKING A GOOD BATH AND YOU SIT ON THE SOFA TO DRINK IT AND SAY THE ROBOTS TOGETHER TO SERVE THE SUPER,THE SAME THING THE THE DINER TIME AND THE BREAKFAST,DON'T THINK ABOUT IF YOU HAVE A FIESTA,THE ROBOTS WILL HIGHLY THE SAME AND MORE AFTER ONE THE ROBOT WILL COME AND SAY ,IT IS 7 O'CLOCK AND THE SUPER IS READY,YOU TALK ONLY AND EVERYTHING ARRIVE ON THE PRESENT MOMENT IN A HAND CLAP TO YOU

fortunegarcia@yahoo.fr

GARCIA FORTUNE'S TECHNOLOGY AND SYTEM

BEFORE THE BEGINNING OF THE SECOND PART OF THE NOVEL STRANGELY WILL PUSH YOU UP TO THE GREAT DOOR OF MY HEART WROTE FOR YOU.IT IS THE ALL INSIDE OF MYSELF WRTTEN THIS ONLY AND SELF BOOK FOR YOU,JUST UNDERSTOOD DURING YOU'RE READING IT,IT IS NOT A BOOK YOU'RE READING,IT IS THE INSIDE OF A MAN FAR AWAY FROM YOU,BUT IN FRONT OF YOU,YOU'RE READING AND THE SECRET YOU'RE READING IS JUST BETWEEN ME AND YOU.YOU'RE MY NOVEL,BUT YOU'RE

LISTENING MY MOUTH SAYING EACH WORDS TO HELP YOU TO UNDERSTAND WHAT I WROTE IF IT IS DIFFICULT FOR YOU TO UNDERSTAND THE BOOK,YOU WILL LISTEN THE SOUND OF MY HAND TYPING EACH WORDS ANOTHER WAY JUST TO HELP YOU WANT ,UNDERSTAND WHAT I WROTE IT IS FOR YOU.AND I WANT NOW IN A SPECIAL DEDICATING TO ALL MY READERS

THIS BOOK IS SPECIALLY DEDICATED TO MY READERS

THANK YOU SO MUCH TO READ ME AND GO ON PURCHASE EACH ONE OF TECHNOLOGIC BOOKS.

I LOVE YOU SO MUCH IN I KEEP YOU IN THE BOTTOM OF MY HEART EVEN IF YOU DON'T SEE ME

FOUNDATION GARCIA FORTUNE

IN THE SECOND PART , I WOULD LIKE TO PRESENT YOU TOO MY BIGGEST WORK SINCE 2005 IN MY

THE FOUNDATION GARCIA FORTUNE

WORKING FOR AND WITH

*   POOR CHILDREN

*   UNDESIRABLE PREGNANT

*   OLD PEOPLE

fortunegarcia@yahoo.fr

CHAPTER 2

THIS BOOK IS WROTE SPECIALLY JUST TO MAKE A LOT CREATION TO BE POSSIBLE

ALL AROUND THE WORLD FOR THE COMPLETE USE OF EACH HUMAN PEOPLE.

BY ALL THESE NEW TECHNOLOGIES AND SYSTEMS,A LOT OF POOR COUNTRIES COULD BECOME FASTLY A GREAT POWER IN A FEW SECOND.

fortunegarcia@yahoo.fr

GARCIA FORTUNE'S TECHNOLOGY AND SYSTEM

NEW TECHNOLOGY AND SYSTEM

I SEE OUR FUTURE COMPLETELY DIFFERENT FROM THE TWENTY FIRST CENTURY.I REMEMBER DURING I WAS A LITTLE BOY STUDENT,THE CHRISTIAN BROTHERS ALWAYS USED TO SAY HE IS LIVING IN THE MOON 90% OF HIS TIME,MY FAMILY TOO ALWAYS SAID THAT THEY CAN'T UNDERSTAND ME,I ALWAYS TALK ABOUT NEW TECHNOLOGY APPROACH I WILL CREATE ALL AROUND MYSELF.MY MOTHER ALWAYS TOLD ME THAT IT IS TOO SOON TO GO SO FAR,I USED TI SAY THAT OUR FUTURE IS RIGHT NOW NOTHING CAN CHANGE IT,WE DON'T HAVE TO WAIT FOR THE FUTURE,WE HAVE TO GO TO THE FUTURE RIGHT NOW AND CREATE EVERYTHING THAT WE HAVE TO CREATE RIGHT NOW.OUR FUTURE CAN HELP OUSELF TO BUILD A NEW TECHNOLOGY WORLD OR LET'S GO DIRECTLY A TECHNOLOGIC COUNTRY WITH MORE PREVENTIONS AND MORE TIME FOR HUMAN PEOPLE WILL HAVE DURING OF THEIR LIFESTYLE GROWTH BY MEDICAL SCIENCE,NEW TECHNOLOGIES AND SYSTEMS ALREADY AROUND US AND WE CAN'T BE BLIND OF THESE NEW CREATIONS NOW THEY ARRIVE ON YOU, US , ME AND ALL OVER THE WORLD .HUMAN PEOPLE CAN MISS ,WE CAN BE SURE OF THAT WE WILL BE ABLE IN THE PLENTY

MIDDLE OF NEW TECHNOLOGIES AND SYSTEMS.FOR THIS REASON,ON A CYBER CAFE OR WEB SPACE AROUND 2005 OR 2006 OR THE LATEST,I SENT A LIST OF NEW CREATIONS ON INTERNET WITH MY NAME INSIDE OF IT,BETWEEN THAT LIST I WROTE TRANPARENT Can,ELECTRONIC BOOKS AND SO ON.I'VE DONE THAT I DIDN'T KNOW HOW I AM GOING TO PUBLISH THOSE CREATIONS AS FAST AS POSSIBLE,I WAIT FOR AFTER THE EARTHQUAKE IN 2010 TO MEET XLIBRIS TO PUBLISH MY 2 FIRST BOOK AND I UNDERSTAND SINCE THIS TIME I COULD BELIEVE IN XLIBRIS,I DECIDE MYSELF TO PUBLISH "100% NEW TECHNOLOGY AND SYSTEM" LIKE MY THIRD BOOK .THERE ARE SO MANY ANOTHERS NEW TECHNOLOGIES WE CAN DEVELOP TO SAVE US AND THE WORLD TO BE COMPLETELY NATURAL AND PRINCIPAL USERS OF FREE ENERGIES OF THE BLUE PLANET,A BOOK WILL BE PUBLISHED IN FRONT OF A PERSON MEMORY .

IN THAT NEW MILLIENIUM,HUMAN PEOPLE BECOME MORE AND MORE FAST STRANGERS TO DISCOVER THE WORLD AND TRAVELLING UNCESSELY.WE APPEAR LIKE IN THE WORLD A MOTHER CARD WHERE SO MANY THINGS ARE GOING TO BE CHANGED.THAT MEANS THE NEEDS OF EVERYBODY GROWTH AND WE ASK OTHER MATERIAL TO ACCOMPLSH OUR LIFETIME WORK.IF AIN A CASE,THERE ARE SO MANY RESEARCHERS IN THE WORLD WHO NEVER FINISHED TO LOOK FOR AND CREATE NEW TECHNOLOGY TO HELP OUR LIFETIME PLEASURE IN WORK LIKE ALSO IN OUR GAMES WHEN WE SAY "NEW TECHNOLOGY AND SYSTEM" FOR US ,WE SEE RIGHT AWAY,NEW CREATION WE ARE GOING TO USE IN OUR FUTURE,THAT'S TRUE. EVERYDAY IN THE WORLD,THAT WAS A NEW CREATION FROM NEW TECHNOLOGY WHICH IS BORN.WHEN YOU SEE IN THE WORLD THAT MEANS ALL AROUND THE WORLD WHATEVER IN ADVANCED COUNTRY OR MENUS ADVANCED COUNTRY.WE CAN SEE THAT IS A GREAT DIFFERENCE IN ADVANCED COUNTRY AND MINUS ADVANCED COUNTRY. IN ADVANCED COUNTRY,WE CAN GOVERNMENT LOOKING SOME HIGH PERFECTION OR PERFECTION IN ALL THEY ARE DOING MUST BE PERFECT FOR RESEARCHERS TO HELP

THEM TO REALIZE THEIR CREATION SOMETIMES FOR PERSONAL USED OR THE GOVERNMENT STRATEGY FOR MILITARY OR SECURITY OF THE COUNTRY SOMETIMES ALSO THAT CAN BE FOR WORLD UTILITY LIKE FOR SOME CREATIONS BY NEW RESEARCHERS THAT CAN BE USED TOO FOR PERSONAL USED BY GOVERNMENT.AND ALSO SOME HOUSE NEEDS PERSONALS NEEDS OF PEOPLE IN THE CASE OF INDUSTRIES.

IN MINUS ADVANCED COUNTRY,THERE ARE SO MANY PEOPLE WHO CREATE NOW TECHNOLOGY AND NEW SYSTEM BY THEIR OWN CAPACITY AND EVERYBODY CAN SEE TOO THE MORE CREATER IN MINUS ADVANCED COUNTRY,THE MORE THEY WILL NEED HOUSE OF PRODUCTION TO PRODUCT GREATLY THEIR NEW TECHNOLOGY.IT'S THE REASON WHY I COMPLETE SOME CREATIONS OF MINE IN THIS BOOK "100% NEW TECHNOLOGY AND SYSTEM" TO REALIZE THEM OR PRODUCE THEM IN OR FOR INDUSTRIES.

fortunegarcia@yahoo.fr

GARCIA FORTUNE'S TECHNOLOGY AND SYSTEM

LIST OF NEW CREATION DEVELOPED

1 * FLYING CAR PART 2

2 * TECHNOLOGY CITY part 2

3 * HARD GLASS DOOR CARS

4 * MODERN AIR PLANES

5 * THE LAZER GLASSES GUNS

6 * THE LAZER COMPUTER

7 * OXYGEN

8 * ELECTRIC MOP

9 * NEW MARKETING SYSTEM

10 * SIMPLE COMPUTER

10 * MY TECHNOLOGIC WATCH TW5

11 * LIGHT AND AUTOMATIC MICROPHONE

12 * NATURAL LAZER GUNS

13 * SUPERVISOR ELECTRIC

14 * HEAD PHONE

15 * MY NEW PERSONAL COMPUTER

16 * MODERN WATER CLOSET

17 * SUN ENERGY

18 * ROBOT LIFE

19 * NATURAL WORLD ENERGY

20 * REGENERATES CELLULARS

21 * FUTURE CARE

22 * GLOBO CAR

23 * UNPLACED PERSONAL COMPUTER

24 * DIRECT SCANER FOR COMPUTER (USING LAZER)

fortunegarcia@yahoo.fr

GARCIA FORTUNE'S TECHNOLOGY AND SYSTEM

1—FLYING CAR PART 2

IN THAT NEW MILLENIUM,ALL CONSTRUSTORS ARE LOOKIING FOR NEW TECHNOLOGY,CREATION MORE EASY TO MAKE WORLD LIFE AS SIMPLE TO WALK AWAY AS POSSIBLE.WHAT TECHNOLOGIST WANT TO BUILD SOMETIMES IS SEEING AND COPYING IN THE NATURE. RIGHT AWAY THE OBJECT CREATED IS FINISHED AND LET'S SEE THE NATURE FORM COMPLETELY.THERE ARE SO MANY YEARS TECHNOLOGIST LOOK FOR HOW TO MAKE A FLYING CAR,YES,I SAID A FLYING CAR SO EASILY.IN THE LAST CENTURIES,TECHNOLOGIST BUILD PLANE,CAR,PHONE,TRAIN,COMPUTERS   SYSTEM,HEL ICOPTER,RADIO,TV,BICYCLE,SPACIAL   PLANE,OTHERS GREATS TECHNOLOGISTS ARE LOOKING FOR TILL THEIR CONTRUCTIONS BETWEEN THEM ROBOT ACTING LIKE HUMAN PEOPLE,FLYING CAR,SPATIAL PLANE ABLE TO TRAVEL WITH MANY PEOPLE THROUGH OUT THE SPACE TO SPACE TO VISIT MOON AS FAST AS POSSIBLE,THE 10 PLANETS IN ALL THE UNIVERS TO HAVE CONTACT WITH OTHERS PEOPLE OF THE UNIVERS WHO CAME EARLY TO SEE THE PLANET EARTH WITH SOMETIMES WANTING TO CONTROL HUMAN OR TO LIVE IN THE WORLD.A FULL MARKETING IS DOING TO ACCEPT THEM WHEN THEY WILL APPEAR ON THEIR TRUE FORM AND FOR THIS REASON ON BAGS,T-SHIRTS AND EVERYTHING ELSE,THEY PREPAIR US,HUMAN PEOPLE TO ACCEPT THEM.THEY ARE SOMETIMES GOOD OR MAD LIKE IN THE WORLD,WE HAVE NICE PEOPLE AND MAD PEOPLE,THE UNIVERS IS LIKE THAT TOO.GOOD EXTRA TERRESTRE HELP THE WORLD TO SAVE

THE PLANET EARTH.MAD EXTRA TERRESTRE OR VISITORS COMING FROM THE UNIVERS JUST COME TO BRING WHAT THEY DON'T HAVE IN THEIR OWN PLANET AND DESTROY HUMAN PEOPLE TOO.PEOPLE AROUND THE WORLD MUST KNOW TO DETECT THEIR BEHAVIOR TO UNKEEP THEM AND SEND THEM AWAY OR KEEP THEM WHEN THEY ARE GOOD OR MAD,SOMETIMES THEY ARE TOO RESCUED FROM THEIR PLANET DESTROYED. THEY HELP US SOMETIMES TO CREATE NEW TECHNOLOGY FOR THE WORLD.A LOT OF CREATIONS WERE SHOWN BY THE VISITORS WHO GAVE US THE WAY TO CREAT THEM.NOW A DAY,WE ESTABLISH CONTACT FIRST WITH VISITORS WHO USED TO COME TO VISIT EARTH SINCE THE CREATION OF THE PLANET EARTH,VISITORS ARE FOLLOWING STEP BY STEP THE EVOLUTION OF THE PLANET EARTH AND ARRIVING IN THE AGE OF HUMAN PEOPLE WHERE THE ATLANTIDE IS THE GREAT FIRST CITY THE MORE DEVELOPED WITH HELP OF VISITORS WITH THE GREATEST CONTACT KEEPED SINCE THE COMPLETE DESTRUCTION OF THE ATLANTIDE BY THE GREATEST CHANGEMENT OF THE PLANET,THE GREATEST EARTHQUAKE DIVIDED THE GREATEST CITY OF THE EARTH IN SO MANY PIECES OF LANDS.THE VISITORS ALWAYS KEEPED CONTACT WITH THE WORLD AND WENT ON THE POWER OF NEW TECHNOLOGY WITH PEOPLE.

NEW TECHNOLOGY IS SOMETIMES EARLY SEEN AND BUILT BY THEM IN THEIR OWN PLANET.EACH ONE HAVE THE SAME BEHAVIOR THE WAY EACH ONE WANT TO KNOW HOW IS MAKING THEIR BODY.

TO KNOW MORE ABOUT A CONSTRUCTION OF A FLYING CAR,DEVELOPING THE TOTAL SUBJECT INSIDE OF THOSE 2 NOVELS AVAILABLE ON

*   HER'S FLYING CAR

fortunegarcia@yahoo.fr

*   THE LAST EAGLE

YOU CAN PURCHASE THEM ON

WWW.AMAZON.COM

TO BUILD A FLYING CAR,IT'S SO SIMPLE THAT ANY ONE COULD KNOW.LET'S SEE THE NATURE WHERE THERE ARE SOME BEASTS WHICH LIVE IN THE AIR AND UNDER THE WATER.THERE IS A CARE TO KEEP THE EGUAL MOMENT IN DIFFERENT POSITION, THE SAME THING CAN BE MADE FOR A FLYING CAR THE WAY TO KEEP AN ONLY GREAT RELATION BETWEEN THE EARTH AND THE AIR. NOTHING IS IMPOSSIBLE TO CREATE IN THE WORLD TO SEE EVERYTHING IMPOSSIBLE,LORD EARLY TOOK THE IMPOSSIBLE AND FOR THE POSSIBLE,THAT MEANT EVERYTHING POSSIBLE.THE FLYING CAR WILL BE A GREAT ADVANTAGE FOR PEOPLE AROUND THE WORLD WHO LIKE TO TRAVEL IN THEIR COUNTRY OR LKE SELF STRANGERS INSIDE THE SAME COUNTRY SO EASILY.THE CAR BY ITSELF IN ONE HOUR CAN BRING ANYONE SO FAR INSIDE THE COUNTRY.DURING THE YEAR,PEOPLE CAN SEE THAT CAR IS VERY UNFASTER TO DRIVE PEOPLE RIGHT AWAY,REALLY RIGHT AWAY SOMEWHERE.NOBODY NEVER UNDERSTAND THAT FLYING THEY ALWAYS WANT IS CERTIFIED BY THE LAW OF NATURE LIKE MANY CREATIONS EARLY CONSTRUCTED. THERE IS ONE THING WHICH MUST BE SEEN BY A FLYING CAR CONSTRUCTOR,THIS IS NATURAL LIFE OF A FROG WHICH LIVES IN WATER AND AIR .THIS IS THE ONLY THING TO SEE GREATLY BY A CONSTRUCTOR.WORKING WITH THE FIRST PART OF LIFE A LIFESTYLE EARLY EXIST TO REALIZE THE FLYING CAR .ANYONE COULD MAKE A DIFFERENT FLYING CAR RESPECTING THE SAME LAW LIKE EYES OF THE WRITER SEES IT IN BACKHOUSE.IT'S SO SIMPLE THAT ANYONE CAN'T IMAGINE IT.ALL ADVICE COULD BE FOUND WITH A GOOD THINKING IN THE NATURE.IF ONE OBJECT FLY,IT COULD RUN.THE SAME THING IN REVERSE CAN SHOW HOW TO CREATE A FLYING CAR.ALL MUST BEGIN SINCE THE CREATION TO ARRIVE IN THAT MILLENIUM. USE EVERY BEASTS,ONE AFTER ONE TO FIND WHICH ONE COULD BE APPOPRIATED TO THE CONTRUCTION OF YOUR

FLYING CAR AND THE SAME TIME,YOU MUST FOLLOW
LAW ALREADY EXISTED.THERE IS NO DOUBT,ANYONE
COULD FIND AND CREATE HIS OWN FLYING CAR LIKE
RESEARCHERS ARE DOING CHINA,JAPON,USA AND OTHERS
COUNTRIES LIKE GERMAN,FRANCE,ITALIE.WHICH ONE IS
GOING TO BUILD OR EARLY BUILD THAT FLYING CAR.THEY
SAY ONE THEM ALREADY CREATED A CAR RUNNING WITH
WATER LIKE OIL.FOR THE MILLENIUM,ALL CONSTRUCTOR
ALL AROUND THE WORLD CONTRUST MODERN CARS
WITH ALL CONTROL SYSTEM OF A COMPUTER.THAT'S
VERY GOOD TO SEE A CAR LIKE TURTLE RUNNING AND
PASSING BY,CAN GO AWAY WITHOUT THINKING TO BE
WORRY ABOUT ANYTHING IS IN THE LITTLE CAR LIKE A
HOUSE WITH ALL THAT HE HAS IN ITSELF.SOME PEOPLE
ARE GOING TO SAY PUT ALL THESE THINGS IN A FLUING
CAR,THAT WILL BE MARVEALLOUS TO FLY AND LAND
DOWN WITH A CAR.I SAY IT AGAIN, THERE IS NOTHING
IMPOSSIBLE,PLEASE UNDERSTAND THAT AND CREATION IS
NOT SO FAR TO SEE A DAY.IMAGINE THAT SOMEONE CAN
HAVE WITHOUT WITH HIS OR HER DEPARTURE WITH
ANYTHING BACK.THAT SAME CAR CAN GO OUT IN THE
SPACE TO VISIT UNIVERS.THAT CAR WITH SOME SPECIAL
NEW INSTRUMENT CAN DO SO EASILY IN THE SPACE AND
RETURN BACK IN THE WORLD WITHOUT PROBLEM AND
THE WAY A CAR SPATIAL STATION IS BUILT IN THE SPACE.
IF SOMEONE LIKE THAT IS SEEING,IT'S MEANT IT CAN BE
DONE SO EASILY TO CROSS ALL PLANETS ONE AFTER ON
VISIT THEM QUIETLY.

I DRIVE MY FLYING CAR IN THE AIR DURING I LAND DOWN
TO ENTER AND SPEND SOMETIMES UNDER THE SEA
TO GO AND SEE BEAUTIFUL FISHES.I GET OUT THE SEA
AND RUN UNTIL MY HOUSE DURING I LISTEN UP A CD .I
ARRIVE UPSIDE MY HOUSE.I USE MY REMOTE CONTROL
TO OPEN THE CEILLING AND GET DOWN MY CAR IN.I
GET DOWN THE CAR TO ENTER IN THE DINING ROOM
,THE FIRST DOOR IS OPENED SUDDENTLY AND I SPEND
IT TO GO IN THE REFRIGERATOR.IT READ MY IDEA AND
SEND ME WHAT I NEED,A GLASS OF JUICE.I TAKE IT AND

WALK TO HAVE A BATH.THE AUTOMATIC BATHROOM PROGRAMMED PREPAIR ITSELF THE BATH TO BE READY FOR MY BATHTIME.I DO IT AND ENTER IN THE DRESSROOM,A ROBOT ON THE PROGRAMMED PLAN OF MY DRESSES FOR THE MONTH,GIVES ME THE DRESSES FOR THE DAY.I SEAT DOWN IN THE LIVING ROOM ON MY DESK TO PREPAIR MY WORK PLAN FOR TOMORROW.AFTER 3 HOURS OF WORK NIGHT,I DECIDE MYSELF TO GO TO SLEEP TO BE READY TO BE READY FOR TOMORROW.I SLEEP ALL NIGHT AND BEGIN MY DAY BY SOME SLOW MUSICS FROM A RADIO STATION.I SING THE SONG I KNOW WHEN THEY SPEND THEM AND SHAKE MY BODY SOMETIMES.I WALK MYSELF THOUGH MY CAR AND OPEN THE CEILLING PLACE TO GET OUT.I DRIVE UNTIL MY JOB STATION.

fortunegarcia@yahoo.fr

GARCIA FORTUNE'S TECHNOLOGY AND SYSTEM

2—TECHNOLOGY CITY PART 2

A CITY IS BUILT.A LOT OF PEOPLE COME AND VISIT IT.THE SUPERB AND MARVEALLOUS KINDS OF HOUSE BUILT IN THIS CITY MAKE US HUMAN PEOPLE REMEMBER OF THE STRANGER CITY THAT THE AZTEC BUILT AND AFTER THEY DISAPPEAR TOGETHER IN THEIR BACK THE CITY. THAT TECHNOLOGIC CITY IS DIFFERENT,EACH HOUSE IS A FLYING HOUSE ABLE TO PROPULSE ITSELF UP AND FLY IN THE CASE OF A GENERAL EARTH CATASTROPH OF WHATEVER IT IS,WAS OR WILL BE.

fortunegarcia@yahoo.fr

GARCIA FORTUNE'S TECHNOLOGY AND SYSTEM

3—HARD GLASS DOOR CAR

THE CAR IS MADE TOTALLY WITH AN ALLIAGE OF IRON AND GLASS WHICH WILL GIVE A COMPLETE HARD GLASS DOOR CAR,THAT MEANS THE COVER OF THE CAR WILL BE MADE TRANSPARENTLY TRANSPARENT.THE DOOR OF THE CAR THAT WE CAN REPLACE THE LATERAL BY A SYSTEM OF A COMPLETE OTHER POSITION WHICH GET UP ENTIRELY IN THE HEAD OF THE CAR.FOR EXAMPLE,WE CAN MAKE IT GET UP IN ALL POSITION STABILIZED ON THE CAP OF THE CAR.THAT WAY ,THE DOOR CAN GET DOWN AND GET UP JUST THE HANDING OF THE BUTTOM POINT.THERE IS ONE KIND WHERE WE BRING UP AND IT STOPS.NOW THE SYSTEM CAN BE APPROUVED ONE WAY COMPLETELY OF HARD GLASS CAN BE GOT UP LIKE THE GLASS IN THE NORMAL DOOR OF A CAR.

fortunegarcia@yahoo.fr

GARCIA FORTUNE'S TECHNOLOGY AND SYSTEM

4—MODERN AIR PLANES

ON OUR DAYS, TRAVEL WITH SECURITY MUST BE A FIX POSITION FOR EVERYBODY WITH ZERO RISK OF ACCIDENT AND WE KNOW AIR PLANES MADE SO SO MANY ACCIDENTS KILLING PEOPLE A GREAT DEAL AND WILL BECOME A GREAT DEAL AND WILL BECOME THE OBJECT OF TERRORIST NOW. WE HAVE TO THINK ABOUT A NEW KIND OF PLANE WITH ALL GREAT SECURITY FLIGHT AIR, WATER FIRE AND AIR CONTROL THE FLIGHT SYSTEM. HERE THEY ARE SOME NEW SYSTEMS TO PROTECT PLANE FROM ACCIDENTS OF ALL KINDS

—SECURITY UNDER THE WATER

—SECURITY ON THE AIR

—SECURITY IN THE FIRE

—SECURITY ON THE WATER

—FLIGHT CONTROL SYSTEMS

—SECURITY INSIDE THE PLANE

—SECURITY OUTSIDE THE PLANE

SECURITY UNDER THE WATER

THE PLANE CAN'T HAVE TIME TO USE ALL SYSTEMS PREPAIRED TO SAVE PEOPLE WHEN THE TROUBLES COME AROUND FASTLY.A NEW SYSTEM TO HELP THE PLANE RUNS AWAY UNDER WATER AND KEEP EVERYBODY ALIVE,SAFE AND WARM.THE PLANE IS UNDER WATER ,WHAT' S HAPPENED?

IT'S HAPPENED THAT EXACTLY

1—ALL DOORS ARE SECURIZED TO MAINTAIN PRESSION OF WATER

2—ALL PARTS OF THE PLANE WHERE AIR CAN ENTER ARE CLOSED TIGHTLY

3—AN ELICE,SPECIALLY 4 HARD FEET IRON TO BE PLACED BEFORE IN THE BACK ON THE PLANE TO CONTROL ITS GET DOWN PLACED FOR THE ACTUAL MOMENT WILL BE DEVELOPPED AND TURNS AROUND TO ACTIVATE THE PLANE UNDER THE WATER

4—A RADAR SYSTEM UNDER THE WATER WILL BE PLACED TO DETECT THE PLANE AND CONTACT THE ACCIDENTAL PLANE BY ANY HELPING BOAT LIKE A SOUS MARIN

SECURITY ON THE AIR

A—WE KNOW EVERY TROUBLES SPEND ON THE AIR LIKE MOTOR DEFECTION,TERRORIST ACT,FIRE AN SO ON.TO CONTROL SOME TROUBLES LIKE THAT ,DEVELOPING A SYSTEM OF SECURITY WHEN THE PLANE HAVE TO FAIL DOWN,A HARD PLASTIC BOAT GROWTH THE SAME TIME ON THE TROUBLE TO ASSURE SECURITYOF PASSENGERS AANYWHERE THE FLIGHT WILL FAIL DOWN

B—IF THE PLANE IS ON FIRE,OXYGEN WILL MISS A GREAT DEAL,AN OXYGEN SYSTEM WAY WILL BW PLACED,EARLY EXISTED A NEW TECHNOLOGY TO GO UNDER WATER

IF FIRE WILL DESTROY COMPLETELY THE PLANE ,A NEW KIND OF CHAIR WITH HARD PLASTIC TO BE GROWTH WHEN PROBLEMS EXISTED,A PARACHUTE CHAIR WITH THE SAME PLASTIC BOAT CAN BE DEVELOPPED

SECURITY IN THE FIRE

IT'S THE CAPACITY FOR THE PLANE TO CONTROL WITH MATERIAL PREPAIRED SPECIALLY FOR THE CASE TO SURVIVE OR TO KEEP EVERYONE ALIVE LIKE HARD PLASTIC BOAT CHAIR OR PARACHUTE AND WITH HARD PLASTIC BOAT CHAIR PLACED TOGETHER

SECURITY ON THE WATER

IT'S POSSIBILITY FOR THE PLANE TO DEVELOP ANY AIR HARD PLASTIC BOAT TO CORRECT ITSELF ON THE WATER

FLIGHT CONTROL SYSTEM

TO CONTROL TERRORIST,A SYSTEM OF DIRECT WAY WILL BE PLACED ON THE PLANE LIKE A MINI COMPUTER TO SEE WHERE PLANE HAVE TO FLYAND WHERE HE HAS TO GET DOWN,NOW, THE GPS SYSTEM IS DONE IT WELL.THE SAME TIME THE SYSTEM AND COMMAND WILL ENTER TO FOLLOW ITS COURSE,ANYBODY CAN'T STOP IT ONLY IF THE AIRPLANE TRAVELLERS CAN GIVE IT TO THE PILOT THE FOLLOWING OR ADVICE TO CORRECT IT LIKE THAT ANY PLANE CAN'T BE MISSED ITS OBJECTIVE

SECURITY INSIDE THE PLANE

ANYTHING CAN SPEND INSIDE THE PLANE,A STRICT SECURITY MUST BE

1—AGAINST GUN

A SYSTEM IDENTIFIED BY A COMPUTER WILL RECEIVE ALL KIND OF GUNS EXISTED WILL BE A SUBJECT OF FEAR FOR PEOPLE INSIDE THE PLANE

2—UNBEHAVIOR PEOPLE

SOMEONE DON'T RESPECT THE FLYING PLANE WAYS AND BEGINS TO UNRESPECT IT. A COMPUTER WILL DETECT IT AND PLACE A LAZER PRISON WHERE HE IS AND YOU GIVE HIM OR HER A QUIET SERUM TO RETURN BACK THE COMPLETE CALM OR THE SAME TIME THE LAZER GUN PRISON CAN STAY TILL THE PLANE ARRIVE AT ITS DESTINATION

3—A BOMB IS PLACED ON, IN,OUT OF THE PLANE,A COMPUTER SYSTEM WILL BE CREATED SPECIALLY FOR THAT CASE IF THE BOMB IS CONNECTED DIRECTLY IN THE MOTOR BOX,THE SYSTEM CONNECTED WITH OTHERS SPECIALLIST OF THE BOMB EFFECT ALL AROUND THE WORLD OR THE WEB OF OTHERS SPECIALLISTS AGENCIES OF THE SUBJECT WILL HAVE INFORMATION TO DISCONNECT THE BOMB OR STBILIZE IT IN ALL THE PART IT IS CONNECTED TO MAKE IT BECOME UNDANGEROUS FOR THE PLANE TRAVELLERS. OTHERS GREATS SYSTEMS CAN BE USED TOO TO DETECT DIRECTLY PROBLEMS IN THE PLANE BY GREAT CONTROL OF COMPUTERS INSIDE AND SPECIALLY ALL AROUND COMPUTERS CONNECTED TO THE ONE OF THE PLANE

SECURITY OUTSIDE THE PLANE

NEW KIND OF PLANES COULD BE GREATED FOR MORE SECURITIES IN TRAVELLING.NEW KINDS OF PLANE COULD BE DEVELOPPED TOO WITH ALL THOSE EXPLANATIONS. WHEN WE TALK ABOUT SECURITIES INSIDE THE PLANE FOR PASSENGERS,A LOT OF VIEWS IN OUR MIND APPEAR

fortunegarcia@yahoo.fr

GARCIA FORTUNE'S TECHNOLOGY AND SYSTEM

5—THE LAZER GLASSES GUNS

SOMETIMES POLICE OFFICERS AND SECRETS AGENTS ARE IN VERY HARD SITUATION AND CAN USE DIRECTLY GUNS BY MOVING. THE GLASSES GUNS ARE CALLING TO BE USED WITHOUT ANY MOVEMENT IF ONLY A LITTLE BOX ON THE HAND TO ACTIVATE IT DIRECTLY BY THINKING THAT MEANS COMMAND DIRECTLY BY IDEA, THE LAZER GLASSES GUN WILL PARALYSE THE PERSON TO KEEP INACTIVE THE PERSON DURING A GREAT DEAL OF TIME TO CAPTIVE THE PERSON. THAT SYSTEM CAN AVOID KILLING OR HURTING THE WANTED. THAT SAME LAZER GLASSES GUN CAN BECOME A GREAT GUN ANTI-LAW CRIMINAL ALLTHE WAY WITH THE COMPANY OF VIEW GLASSES ANYWHERE OR ULTRA VIOLET GLASSES WITH HEADPHONE DIRECTLY CONNECTED ON HEAR TO HEAR EVERYTHING EVERYWHERE AROUND THE PLACE WE ARE STABILIZED.

fortunegarcia@yahoo.fr

GARCIA FORTUNE'S TECHNOLOGY AND SYSTEM

6—THE LAZER COMPUTER

THERE ARE SO MANY GUNS AROUND THE WORLD WHICH ARE CREATED JUST TO KILL.SOMETIMES GENERAL ORGANIZATION OF POLICE SURELY NEVER NEED OR DON'T NEED TO THINK ABOUT TO KILL BUT TO CAPTURE.FOR A LONG NOW,WE LOOK FOR LIKE RESEARCHERS OF NEW GUNS TO HELP DIFFERENTS DEPARTEMENT OF POLICE TO STAY IN WHAT THEY ARE DOING AND WHAT THEY REALLY WANT.THEY ARE SO MANY RESEARCHERS TOO WHO DEVELOP NEW KINDS OF GUNS AND FINAL OPPORTUNITY FROM GOVERNMENT TO CREATE CERTAIN GUNS.SO MANY STAY TOO AND FOREVER UNCONSTRUCTED ABLE JUST FOR THE REASONS NOBODY SEE THEIR NEEDS EXISTENCE.IT'S THE REASON WHY SO MANY OTHERS ORGANIZATIONS FINANCE THESE NEW CREATIONS TO CREATE THEIR GUNS TO USE THEM SOMETIMES WITHOUT ANY PERMITS FROM ANY COUNTRY.THESE GUNS BECOMES DIRECTLY UNDER LAWS FOR THEIR CREATORS.SOME NEW GOVERNMENT,WHEN THEY KNOW WHO AND WHERE,HOW CERTAINS KINDS OF GREATS GUNS ARE VERY DANGEROUSLY DANGEROUS FOR THE WORLD,THEY TRY RIGHT AWAY TO DESTROY THEM AND THEIR HOUSE OF CONSTRUCTION.BETWEEN ALL GUNS TO CAPTURE UNLAW PEOPLE,THERE IS ONE WE CAN CREATE WITHOUT ANY PROBLEMS OR DANGEROUS DESEASES FOR ANYONE.EVEN FOR PEOPLE WHO WILL BE CAPTURED.IT'S A GUN WITH AN ENERGY OR WE COULD SAY ETERNAL ENERGY OR ONE 1 ENERGY WE CAN REPLACE WHEN IT IS PLACED IN THE GUN.THE CREATION WILL COST MINUS DEPENSES FOR

DEPARTMENT OF CONSTRUCTION AND WILL BE EASIER TO USE BY DEPARTEMENT OF POLICE WITH A NET INCOMEFOR MATERIALS GUNS.DEPEND ON AN ONLY ENERGY WHICH IS AUTO CREATED BY ITSELF INSIDE THE GUN,THAT NEEDS NO NEED FOR MATERIALS GUNS TO CREATE.THE GUN CAN TAKE DIFFERENT FORM BY THE CONSTRUCTION DEPEND ON TOO WHAT HE WANTS TO LIGHT SOMETHING,LIGHT TO BRING A HEAVY BRIGHTNESS FOR LONG DISTANCE USING THERE WILL NO PROBLEM AND CAN BE SEEN BY ANY CONSTRUCTORS ON PROTOTYPE LIKE WE REALLY TALK ABOUT IT IN USE AND IN CREATION.THAT LAST GUN WILL ASK WORKS ABOUT THE ENERGY WILL BE IN IT.THAT ENERGY MUST BE UNENRICHED FROM GREATS CAPACITIES TO POSSESS NOW A DAY TO KILL AND OBSERVE INSIDE AND OUTSIDE.NO WAY,AFTER TO RESOLVE THAT FIRST PART,A LITTLE GADGET IS GOING TO BE CREATED ,CONTROLLED THE FORM OF THAT ENERGY,SOMEONE WOULD NEED,WOULD LIKE TO GIVE IT TO CAPTURE IT IN THE FIRST STEP,IN SECOND TO CREAT OR WRITE A BILLBOARD IN THE MARKETING SYSTEM.THAT WRITING WILL BE KEPT FOREVER OR FOR A DURING PEOPLE CAN CONTROL IT.THE GREAT PART IS IT WILL BE WITHOUT DANGEROUS IF THAT BILLBOARD IS PLACED SOMEWHERE,EVERYONE CAN TOUCH IT TO AVOID BURNING OR GREAT BURNS.BY THE SAME TIME CERTAIN WORKERS ARE GOING TO USE THAT ENERGY TO WRITE OR TO CREATE PROTOTYPE OF ANYTHING THEY WANT.TO RETURN BACK IN THE GUN,IT WILL BE LIKE A NORMAL GUN TO CONTROL ENERGYTHAT WILL BE USED IN NEED.THAT UNENRICHED LAZER GUN IN SHOOTING WILL BE LIKE A SPIRAL TO CAPTURE THE PERSON ONLY IT NEEDS WITH A GREAT VISUAL TARGET POINT THE ONE THE DEPARTMENT OF POLICE WILL NEED AND WANT TO KEEP HIM IN PLACE DURING A LONG TIME JUST TO USE LAW OF POLICE.THAT LAZER GUN CAN'T KILL NOR DESTROY THE BODY OF SOMEONE.IT WILL BE JUST TO CAPTIVE AN UNLAW PEOPLE GREAT OR UNGREAT UNRESPECT OF THE LAW.THE CAN GUN CAN KEEP ONLY SOMEONE QUIET IN A BAD ACTION TOO.IT WILL BE LIKE A NATURAL PRISON WITHOUT ANY TO FLY AWAY.THE ONLY KEY WILL

STAY IN THE GUN.BETWEEN ALL DESTRUCTIONS,GUNS WHICH EXISTED IN THE WORLD CAN BE DESTROYED TO BE REPLACED BY NATURAL GUNS.THEIR CAPACITY CAN DESTROY THE WORLD AND HUMAN PEOPLE AROUND IT.TODAY THE WORLD NEEDS NEW CREATIONS,NEW CREATIONS COULD HELP IT AND PEOPLE AROUND IT.THE GUN HUMAN PEOPLE CONSTRUST SINCE THE BEGINNING OF THE WORLD IS TO DESTROY IT WITH US,HUMAN PEOPLE EXISTENCE OF THE WORLD DON'T ASK THAT.IN IT CREATION,EVERYTHING,YES, EVERYTHING ISADDED TO THE WORLD FOR OUR USES OR NEEDS.LOST THE CAPACITY OF THE WORLD IN THE UNIVERSAL CO-LIVED IN UNITY AND ALSO AUTO BUILT,WE DON'T TAKE TIME TO FOLLOW THEM AND USE THEM.WE GET OUT FROM THAT UNITY,CREATE OUR OWN DESTRUCTION TECHNONLOGY TO DESTRUCT THE WORLD.OUR TECHNONLOGY DON'T FOLLOW THE LAW OF THE UNIVERSAL IN CONTACT WITH THE WORLD AND UNIVERSAL APPLICATION.THAT LAW IS:

EVRYTHING CREATED FOR THE WORLD RETURN BACK TO THE NATURE,THE NATURE FOR THE WORLD ,THE WORLD FOR THE UNIVERS.AT THE CONTRARY,WE USE WHAT WORLD HAVE TO CREATE NEW TECHNOLOGY TO DESTROY THE WORLD AND HUMAN PEOPLE WITH IT.THE WORLD SURELY REPAIR SOMETIMES BY HELP OF THE UNIVERS IT OWNS PROBLEMS. THERE ARE SOME IT WOULD TAKE IT SO MANY TIME OR TILL THE DESTRUCTION OF THE WORLD TO RESULT LIKE WARM AND COLD PROBLEM OF THE WORLD,IT WILL NEVER BE RESOLVED,ONLY IF WE STOP USING BAD REJECTION IN THE AIR,IN OUR MOTHER NATURE.WE CAN SAVE AGAIN THE WORLD BY FIRST OUR NEW CREATION TO REPLACE THE ONES THERE BEFORE.EVERYBODY KNOWS IT WOULD BE EASIER TO DO BY A CERTAIN OPPOSITION OF CERTAIN MANAGERS OR CERTAIN COUNTRIES.THEY SEE THAT WILL COST US MONEY OF THE ALL WORLD AND IT WILL BE IMPOSSIBLE TO DO SO.WE CAN SEE TOO,THE MORE WE TAKE TIME FOR A PLAN TO CONTROL OUR BAD BEHAVIOR OF THEORY OF REJECT WE POSSESS SINCE THE FIRST TIME HUMAN PEOPLE IS CREATED THE FIRST

THING,WE WILL NEVER BE ANYWHERE THAN BUILD NEW CREATION OR NEW GREATS DANGEROUS DESTRUCTION AGAINST THE WORLD AND US.THE WORLD IS OUR HOUSE. IF WE DESTROY IT ,ONE DAY LIKE WE EXPLORE MARS IN THAT CENTURY,WE COULD THINK A PLANET,THE PLANET WITHOUT OXYGEN COULD RECEIVE US IN IT.WHEN THE WORLD WILL BE IN ITS LAST DAY OR TIME,WE WILL NEED TO GO AWAY IN THE UNIVERS FROM DESTRUCTED WORLD TO ANOTHER PLANET,THATWILL BE THE SAME EVERYTIME WE WILL FINISH TO BUILD,TO DESTROY AFTER AND LOOK FOR TO GO AWAY TO BE ALIVE. TO SAVE OUR LAST ,THE WAY YOU CAN UNDERSTAND RIGHT NOW THE WORLD IS FOR YOU,YOU FOR THE WORLD,YOU WILL HAVE A FEELING TO SAVE OUR LAST PLANET,ONLY HOUSE WHICH NEVER EXISTED IN THE UNIVERS POWER.YOU HAVE TO KNOW TOO,THE UNIVERS IS MADE ONE WITH THE WORLD,ONE WITH US ,ONE WITH OTHERS PLANETS.THE WAY ONE COULD BE SAVED,THE LAST THEORY OF THE UNIVERS WILL BE APPLICATED FOR EVERY CASE OF ANY PLANETS.ALL PLANETS ARE IN CONTACT WITH ONE TO ANOTHER LIKE A GREAT CIRCLE'S TECHNOLOGY AND SYSTEM

fortunegarcia@yahoo.fr

GARCIA FORTUNE'S TECHNOLOGY AND SYSTEM

7—OXYGEN

FOR THE FUTURE,OXYGEN WILL SOMETHING MISSING,WE DON'T KNOW TRUELY WHAT WILL BE THE PRINCIPAL OBJECT OF THAT MISSING IF IT IS NOT THE WARM AND COLD WILL AFFECT THAT TIME.TO CREATE OXYGEN,A NEW KIND OF MATERIAL IS CREATED TO PRODUCE ITS OWN OXYGEN BY ALL KIND OF GAS.

THERE ARE SOME SICKNESS IN THE WORLD NEED TO BE TREATED WITH OXYGEN.LIKE SOMETIMES SOME PEOPLE FALL DOWN AND LOST THEIR MEMORY.THE RECENT RESEARCH WE DO SHOW THAT IF WE SEND OXYGEN BY NOSE TO THE HEAD ,THE PERSON WILL RETURN BACK NORMAL STATEMENT IN FEW MINUTES .THE RESEARCH WE DO SHOW THAT IN THE MOMENT OF DIFFERENT KIND OF CRISIS,THE BRAIN DON'T RECEIVE QUITE OXYGEN TO DISTRIBUATE IN THE HEAD THAT 'LL CAUSE IN PARTICULARITY CRISIS.NOW,A NEW KIND OF CARE MUST BE CREATED AND USED WITH OXYGEN SPECIALLY FOR CRISIS PEOPLE,THERE ARE SO MANY OTHERS SICKNESS WILL BE ABLE TO BE TREATED BY OXYGEN AND SOME SUPPLEMENT OF NATURAL AIR.

fortunegarcia@yahoo.fr

GARCIA FORTUNE'S TECHNOLOGY AND SYSTEM

8—ELECTRIC MOP

THE MOP LIKE WE KNOW IT COULD BE MODERNIZE WITH ELECTRICITY,SOMEWHERE SPECIAL TO KEEP THE WATER AND RESEND IT AGAIN AFTER A FIRST WASH.

GENERALLY WE USE IT TO CLEAN HOUSE .IT KEEPS WATER AND PEOPLE HAVE TO EMPTY THAT WATER AND AFTER ALL TO FULL IT AGAIN IN ITS HAIR.NOW WE CANTALK ABOUT THE ELECTRIC MOP WITH A LONG WOOD AND A POCKET TO HELP WATER IN A TURNING HAIR WATER TO EMPTY THE WATER AND CLOSE ON ITS HAIR WHEN IT 'S NOT IN NEED.

fortunegarcia@yahoo.fr

GARCIA FORTUNE'S TECHNOLOGY AND SYSTEM

9—NEW MARKETING SYSTEM

THE MARKETING SYSTEM OF THE FUTURE WILL CHANGE COMPLETELY BY A MODERN CREATION SPECIALLY MADE FOR THE MARKETING ALL AROUND THE WORLD.

PERSONAL ELECTRONIC THEORY

DURING THE LAST CENTURY ,BUSINESS PEOPLE TRY TO MAKE GROWTH THEIR ANNUAL INCOME TO ADJUST THEIR CAPITAL BY SELLING MORE,THE MOST THEY COULD THEIR MERCHANDISES.BETWEEN ALL THAT ,THERE IS MARKETING WHICH PLAYS A REAL PART TO SELL STOCK OF MERCHANDISES MORE AND MORE INVESTORS INVEST IN INDUSTRY OF PRODUCTION TO CONTRUCT MORE PRODUCTS.THE DIFFERENCE IS HIGH ALL OVER THE EACH PRODUCTS DISTRIBUTION HOUSE SEND ON THE MARKET.BUSINESS PEOPLE NEVER FINISHED TO FIGHT TO KEEP THE FIRST PLACE ON THE MARKET PLACE WITH THEIR PRODUCTS.WHICH MUST BE VERY AND VERY WELL REPRESENTED BY THE FACE OF THE PRODUCT PLAN.THE PRESENTATION OF THE PRODUCT IS THE EFFECT OF THE PRODUCT TO HELP PEOPLE TO SEE IT AND BUY IT THAT WE CAN CALL INDIRECT MARKETING PLAN SELLING .

NOW LET'S SEE WHAT THE DIRECT MARKETING PLAN CAN DO WITH A BEST FIGHT ,A DIRECT FIGHT WHERE PEOPLE CUSTOMERS CHOOSE THE PRODUCT THEY KNOW ONLY BY HEARING AND KNOWLEDGE OF THE LAST NEW

PRODUCT THEY KNOW.DIRECT MARKETING PLAN IS DOING NOW ,WE CAN DIRECTLY ON THE MARKET PLACE BUY IT WITH NO PROBLEM OUT OF OUR MIND THAT MEANT NO UNCONTROLLED SOLD ,NO BAD INCOME. WITH EVERYTHING WE KNOW IN MARKETING PLAN,THE CREATION OF NEW SYSTEM TO HELP CUSTOMERS TO BUY AND PAY FASTLY.

WE SEE TO TALK ABOUT THAT A PRODUCT WHICH CAN TALK TO ITS CUSTOMERS BY TRUE VOICE INCORPORATED IN THE PRODUCT BY A MINI SYSTEM WHO WOULD LIKE TO BUY IT FASTLY IF THE CUSTOMER IS NEED OF THE PRODUCT .

IMAGINE YOU APPEAR IN A MARKET PLACE TO BUY A SOAP. YOU STAY FACE TO FACE TO ONE PRODUCT,ONE SOAP IMMEDIATLY THE SOAP YOU SEE EXPLAINS ITSELF LIKE THAT FOR EXAMPLE

—I AM A BEAUTIFUL SOAP, I CAN CLEAN YOUR SKIN EASILY AND HELP YOU TO FEEL FRESH DURING ANY NICE DAY,BUY ME TO PROTECT YOURSELF

THA MEANT THE PRODUCT SPEACK DIRECTLY TO ITS CUSTOMER AND HELP THE SAME TIME AND IMMEDIATELY THE CUSTOMER TO CHOOSE AND BUY FASTLY WHAT HE OR SHE REALLY NEED.

THE NEW MARKETING PLAN SYSTEM ABOUT THAT MEANS :

WITH A PERFECT PRESENTATION OF THE PRODUCT

WE ADMIT A NEW SYSTEM OF SOUND TO MAKE THE PRODUCT TALKS ABOUT ITS BEST POSSIBILITY TO ITS CUSTOMERS.IT'S VERY SIMPLE THE WAY TO KEEP THE USUAL WRITING OF HOW TO USE THE PRODUCT FACE TO FACE TO THE CUSTOMERS.LIKE TOYS ,WE INSTALL A SYSTEM OF SOUND ,AN AUTOMATIC SYSTEM OF SONG BY TOUCHING THE PRODUCT JUST BY HOT FEELING.THE

PRODUCT AUTOMATICALLY BEGINS TO TALK AND SPEAK TO ITS CUSTOMERS.

THE LAST SYSTEM OF MARKETING COULD GROWTH YOUR ANNUAL INCOME IN A PRODUCTION ENTERPRISE THE WAY THE PRODUCT IS DIRECTLY FACE TO FACE TO THE CUSTOMERS.THAT NEW SYSTEM IS NOT THE DEEREST TO ADJUST IT ON A PRODUCT LIKE CELLULAR TOYS AND EVERY TOYS WHICH TALK TO USERS.

THE SYSTEM IS SIMPLE

1—CAN PLACE THE SOUND SYSTEM INSIDE THE COVER OF THE PRODUCT BOX FOR PRODUCTS WHICH DON'T POSSESS COVER BOX,A SYSTEM CLEAR OF VOICE CAN BE PLACED ON THE CAP

2—ANY KIND OF PRODUCT ,WHATEVER IT IS ,CAN USE THE THAT SYSTEM,JUST TO SEE WHERE YOU CAN PLACE ON THE PRODUCT BY YOUR OWN DESIRE THE SOUND VOICE AND HELP THE PRODUCT,YOU PRODUCT SELIING GROWTH.

fortunegarcia@yahoo.fr

GARCIA FORTUNE'S TECHNOLOGY AND SYSTEM

10—SIMPLE COMPUTER

OUR 21ST CENTURY CAN REPLACE EVERYTHING WHICH CAN BECOME ONE WITH IT. OUR VIEW IT IT CAN CONTROL EVERYTHING FACE TO FACE OR IN OUR BACK WITHOUT ANY TROUBLE .OUR HOUSES,CAR CLOTHES,MEALS,CHILDREN CARES,FROM THE SIMPLEST TO THE GREATEST IN THE HOUSE.THE IMPORTANT IS TO CONTROL THAT MACHINE WITH GREAT CAPACITY TO HELPIT TO DO WELL LIKE WE WANT IT OR TO DO IT IN OUR LIFETIME.OUR TV,WATCH,RADIO,MOBILE GSM WILL BE ONE,WHAT IS ALREADY DONE WHEN I LOOK AT THE DATE I WROTE THAT DATA IN 1990,IT'S FAR AWAY FROM TODAY.ALL THESE MATERIALS WE CAN USE THEM IN ONE WHEREVER WE ARE .THE FIRST TIME VIEW IS TO HAVE INFORMATION AND MAKE CONVERSATION WITH ANYBODY TO USE INFORMATION WE NEED TO MAKE UP ONE'S MIND. THERE WILL BE NO PROBLEM TO CHANGE STEP BY STEP ONE TO ONE TO USE THAT MINI SIMPLE COMPUTER WHICH CAN BE CONNECTED DIRECTLY ON THE WEB.THERE ARE SOME PEOPLE WHO CAN THINK THAT WILL BE GREAT TO USE THESE 5 HOME MATERIALS AND PORTABLE TOO FOR THE GSM OUT JUST TO USE THEM ANYWHERE QUIETLY FOR EXTRA CHANGE ON A PUBLIC PLACE YOU LOOK AT TV AROUND NEWS OR A FILM YOU CHOOSE ON THE WEB. PHONE SOMEONE AND USE INFORMATION THE SAME FROM COMPUTER YOUR ONLY MINI COMPUTER WHERE YOU CAN PREPAIR ,ARRANGE ,COMMAND FROM WHERE YOU ARE MATERIALS HOUSE SOMETHING YOU FORGET TO DO BEFORE YOU LEAVE THE HOUSE.OUR NEED TO BE SIMPLE

IN EVERYTHING ,FOR THAT NEW CENTURY IS THERE,WE CAN BLIND IT.JUST WE NEED TOO MORE TO THINK ABOUT WHAT WE NEED MORE CONFORTABLE OR MORE EASILY TO DO RIGHT AWAY TO HAVE IT FOR EXAMPLE,THERE ARE SOME WISE POINT ,WE CAN TRANSLATETO ANOTHER LIKE,PUBLISHING HOUSE,AUTO CONSTRUCTION,HOUSE KE EPER,CONTRUCTOR,BODYGUARDS,HOUSE CONTRUCTIONS AND EVERYTHING HEAVY WE CAN BE MANAGED ONLY BY COMPUTER OR COMPLETE ROBOT OR HUMANOID THAT MEANT ROBOT AND HUMAN THE SAME TIME WORKER.WE BEGIN TO REPLACE COMPUTER IN CERTAIN CONSTRUCTION.THE IMPORTANT NOW IT TO INSTALL IT COMPLETELY AND WE CAN CONTROL IT BY ONLY BY OUR SIMPLE COMPUTER.LET'SLOOK AT SOMETHING AROUND PUBLISHING HOUSE WITH ELECTRONIC BOOK WHICH COMES AND ENTERS IN A COMPUTER WHILE E-BOOK IS EDITED OR PRINTED TO COMPLETE IT.THE BOOK WILL BE NOW READ OR LISTEN BY US.

TO RETURN BACK TO OUR SIMPLE COMPUTER WE CAN CREATE IT AS EASY AS SIMPLE WE CAN SEE IT.

fortunegarcia@yahoo.fr

GARCIA FORTUNE'S TECHNOLOGY AND SYSTEM

10 A—MY TECHNOLOGIC WATCH

THESE 5 MATERIALS WILL BE IN ONE ONLY IN A SIMPLE
WATCH OR SO SIMPLE CALLING THECHNOLOGIC WATCH.
TO CREATE IT EVERYTHING WILL BE IN MINIATURE OR
SO LITTLE THAT WE CAN TRANSFER OUR COMMAND
ONE TO ANOTHER AND ANOTHER TO ONE.LIKE GSM
PHONE , IN THE TECH WATCH, A PLACE WILL BE RESERVED
FOR MANUAL PART TOUCHED TO ENTER COMMAND
OR DIRECTLY BY VOICE.THAT MANUAL PART WILL
RESERVR THE RIGHT TOO FOR ALL MATERIALS LIKE
TV,WATCH,RADIO,MOBILE PHONE,COMPUTER IN THE
SAME WATCH COULD SEND LAZER PICTURES TO WORK ON
THE VISUAL FALSE COMPUTER AND THE SAME THING TOO
FOR THE REST OF EACH ONE OF THE FIVE 5 MATERIALS
AND MORE INCLUDED IN THE TECH WATCH OBJECTS AND
CONTROL THEM.IT CAN BE USED ANYWHERE AND WHERE
ANYBODY WANT LIKE UNDER THE SEA ,SPACE AND OTHERS
PLANETS.THE IMPORTANT IS TO HAVE THESE NEEDS IN
FRONT OF US JUST WHEN WE NEED TO USE THEM WHEN
NEED IS NEED WHERE WE DON'T NEED TO ENTER HOME
TO USE TV AND THE REST OF 5 THINGS OF THE SPECIAL
WATCH.WE BRING EVERYWHERE A LISTEN RADIO OR A
LITTLE RADIO ANYWHERE WE GO,OUR GSM,EVERY 5 NEEDS
WILL BE IN ONE LITTLE WATCH ON OUR HAND WITH
PROBLEM TO CARE ABOUT INSIDE AND OUTSIDE THE
HOUSE,OUR CHILDREN WITH THE SAME WATCH WITH A
GPS COULD BE KNOWN WHERE THEY ARE BY IMMEDIATLY
BY YOU BY ASKING THE NEED.THAT WE WILL USE WITH ANY

COMPLICATIONS LIKE WE USE ANY OF THOSE MATERIALS
LONELY.

THE FUTURE WILL BE OURS

GARCIA FORTUNE'S TECHNOLOGY AND SYSTEM

11—LIGHT AND AUTOMATIC MICROPHONE

EVERYWHERE AND EVERY CEREMONY,WE USE A MICROPHONE .ORDINARY MICRO IS SOMETIMES WITH PLOG TOTAKE CONTACT IN OTHER INSTRUMENT.NOW,WE CAN TALK ABOUT A NEW MICRO WHERE SOUND WILL BE ALWAYS SEND ON INSTRUMENT.IT WILL BE SIMPLE OR LITTLE LIKE A REMOTE CONTROL.LIGHT WILL BE ADDED WHERE VOICE IS ENTERED IN THE MICRO.A LITTLE LIGHT OF SEVERAL COLORS .THE VOICE CAN BE CONTROLLED BY A REGULATOR ON THE SAME MINI MICRO.

GARCIA FORTUNE'S TECHNOLOGY AND SYSTEM

12—NATURAL LAZER GUN

13—SUPERVISOR ELECTRIC

DURING THESE DAYS,THERE ARE SO MANY NEW ELECTRIC CREATION.WHEN THEY RECEIVE ELECTRICITY SOMETIMES,THEY RECEIVE THE SAME TIME ELECTRIC SHOCK,NOW THERE ARE A NEW KIND OF CREATION WE THINK ABOUT LIKE WHEN ELECTRICITY WITH GREAT POWER ENTERS IN A NEW CREATION OR OLD CREATION,THERE IS A SYSTEM INSIDE THE CREATION WHICH CAN ACCEPT OR REFUSE ELECTRICITY TO POWER IT.WE THINK ABOUT A SYSTEM OF CONTROL LIKE A MINI COMPUTER INSIDE THE CREATION LIKE A BRAIN,YES,LIKE THE BRAIN OF THE CREATION CAN REALIZE IF IT'S OKEY OR NOT TO POWER OR SEND ONE ELECTRICITY OUT OF THE CREATION. LIKE THAT,SO MANY CONSTRUCTION CAN EVEN IF THE ADVICE COMPANY GIVE TO PROTECT IT .WHATEVER IT TAKES SO MANY HOUSEWARE ARE SHOCKED JUST WHEN A BAD ELECTRICITY ENTERS INSIDE IT.THAT WILL HELP TO SAVE AND HELP CONTROL SO MANY HOUSEWARE,ABOUT THAT ,WE CAN ENTER SOMETHING ELSE INSIDE THE NEW HOUSEWARE TO APPEAR WITH A NEW KIND OF SYSTEM ,NO WAY,WHATEVER THE SUBJECT WE USE TO SAVE ALL AROUD ELECTRIC SHOCK,WE CAN BE THE FIRST WINER.

I WROTE THIS MARCH 23RD,2004

fortunegarcia@yahoo.fr

GARCIA FORTUNE'S TECHNOLOGY AND SYSTEM

14—HEAD PHONE

A NEW KIND OF HEAD PHONE WILL REPLACE TOTALLY THE ABILITY OF THE PHONE.THE HEAD PHONE WILL BE TOTALLY THE THE PERSONAL PHONE WITH NOTHING ELSE IN OUR HAND WITH ALL ABILITY TO DO WHAT A GSM PHONE COULD DO WITH MORE KEEPING CONTACT WITH OUR DIRECT PERSONAL HOME COMPUTER KEEPER TO KNOW IF EVERYTRHING IS WORKING WELL AND SPEND ANOTHER NEEDS OR ASKS WE ARE FORGOTTEN BEFORE LEAVING THE HOUSE.

GARCIA FORTUNE'S TECHNOLOGY AND SYSTEM

15—MY NEW PERSONAL COMPUTER

IT IS ABLE TO DO EVERYTHING

1) TO TALK TO ME BY ITS PERSONAL SELF INDEPENDENT MIND ABILITY

2) TO WRITE,TO SEE AND LISTEN BY ITSELF OWNLY THE CONVERSATION,FAX,SCANIZE AND COPY THE SAME TIME BY HELP OF ITS EYES COULD BE MORE DEVELOPED FROM THE CAMERA WITH ALL ABILITY AUTOMATICALLY TO REALIZE BY ITSELF ALL WE NEED WITHOUT A HUMAN BEING TOUCHING IT IF ONLY WE REALLY WANT TO TOUCH IT

3) TO CONNECT ITSELF IN INTERNET WITHOUT LINE AND GIVE ME ALL INFORMATIONS ,ADVICES,I NEED ABOUT SOMETHING DURING IT IS DIRECTLY CONNECTED ON MY PHONE ,MY CELLULAR PHONE,WHERE WE ARE TALKING TOGETHER LIKE TWO PERSONS

—TO CARE OF ITSELF THAT MEANS TO CONTROL ITS SYSTEM OR ITS DEFECTIONS SYSTEM IN SOFTWARE AND HARDWARE TO BE ABLE TO SEE AND SAY WHERE THE PROBLEM IS

—TO CONTROL MY HOUSE,LIKE A ROBOT,A COMPLETE ROBOT ASSOCIATED TO ALL ROBOTS IN THE HOUSE TO BE THE FIRST COMMANDER TOTALLY AND MY CAR EVERYTHING I'M DOING AND AUTOMATICALLY GIVES RESULTS OR RESOLVATIONS

OR ADVICE ABOUT ANYTHING ALL AROUND THE HOUSE,ME OUTSIDE THE HOUSE TELLING ME WHERE I CAN SPEND WITH MY CAR TO GO FASTLY,MY WORK WHERE I AM RIGHT NOW,THAT MEANS MY OWN PERSONAL COMPUTER INSIDE MY HOUSE IS ABLE TO DO A LOT OF THINGS INSIDE THE HOUSE,OUTSIDE THE HOUSE FOR ME ,MY CHILDREN AND MY WIFE.

MY NEW PERSONAL COMPUTER OR MY LAPTOP IS COMPLETELY INDEPENDENT IN THE IT IS PERSONALIZE FROM MY CARACTER AND POINT OF VIEW WHERE IT CAN PARTICIPATE LIKE A DOG IN ALL MY ACTIVITIES WITHOUT ASKING IT BY VOICE EVERYTHING IS DONE DIRECTLY ON TIME WE ARE TALKING TO THE PC OR PL TO REALIZE ALL OUR PROJECTS DEPENDING ON ITS OWN GREAT CAPACITY TO CARE OF OUR NEEDS.A SPECIAL SYSTEM WILL BE DEVELOPED FOR BLINDS PEOPLE AND FOR UNTALKING PEOPLE,THIS PC,PL COULD BE FOUND AT BETTER PRICE. TODAY IN A POOR COUNTRY WHERE NECESITY ON NEEDS OF THOSE MATERIALS ARE REAL COST MORE AND MORE HIGH.THE PLACE OF THE PC,PL ARE VERY IMPORTANT IN THOSE POORS COUNTRIES TO HELP THEM CREATE MORE A BETTER WAY OF LIFE ALL AROUND INTERNET WHERE THE RISK OF WORKING IS HIGH THAN DIRECTLY IN THE SOCIETY WHERE WE ARE LIVING IN THE VIRTUAL ENTERPRISE TO THE VIRTUAL ENTERPRISE

OUR PERSONAL COMPUTER WILL ABLE TO BE THE SAME TIME TO BE IN WHAT EVER WE WANT IT IN THE HOUSE OR OUTSIDE TO PROJECT VIRTUAL LAZER DRAWING TV, LAPTOP, E-BOOK AND ALL NEW TECHNOLOGY WILL BE CREATE WITH A SCREEN THAT MEAN THE TV AND EVERYTHING HAVING A SCREEN WILL BE VIRTUAL AND CONTROLLED DIRECTLY BY THE PERSONAL HOME COMPUTER AND WE CONTROL BY VOICE DIRECTLY OR A SPECIAL REMOTE CONTROL CONTROLLING MADE A CREATED SPECIALLY FOR VIRTUAL CREATIONS.

A NEW WAY OF LIFE OF TECHNOLOGY IS WAITING FOR US

fortunegarcia@yahoo.fr

GARCIA FORTUNE'S TECHNOLOGY AND SYSTEM

NEW CREATION

FOR POOR COUNTRIES,A PLACE WILL CREATE WHERE EVERYBODY SURELY THE MORE,THE YOUNGEST PEOPLE WE WILL UPGRADE MORE ACCESS TO THE WEB COME AND WORK WITH A VERY LOW PRICE TO OPEN THE VIRTUAL STUDYING TO A HIGHER QUANTITY OF PEOPLE OR YOUNGEST PEOPLE.THIS WILL BE THE WORLD PERSONAL OF WORKING PEOPLE TO BE GRANTED AND GET DOWN THE RISK OF UNWORK PEOPLE ALL AROUND THE COUNTRY AND A LOT OF COUNTRIES COULD FIND THE SAME PROCESSUS AROUND THOSE POORS COUNTRIES. TODAY ,THE WORLD NEW ACCESS OF TECHNOLOGY MORE ADVANCED OF NEW PERSONAL COMPUTER AND NEW PERSONAL LAPTOP WILL BRING TO THE WORLD UP AND SAVE MORE ALL AROUND US HUMAN BEING.

THE RISK TO CONTRO US DEFINITELY AND TOTALLY BY ITS OWN ABILITY FREE WILL BE CONTROLLED BY US HUMAN PEOPLE.THE FREEDOM OF PC OR PL WILL NOT BE AN INDEPENDENCE TOTAL,IT WILL SET ON OUR STOPPING POINT OF VIEW,I MEAN THE PC OR PLWILL HAVE A DEFINITE POINT OF STOP, A LIMIT,A HIGH POIT OF LIMIT COULD STOP IT ON CONTROL BY US,HUMAN PEOPLE.IT'S VERY IMPORTANT TO SET THIS LAST PART CAN HELP THE MACHINE TO TAKE A TOTAL FREEDOM And be AWAY FROM ALL OUR COMMAND.

fortunegarcia@yahoo.fre

GARCIA FORTUNE'S TECHNOLOGY AND SYSTEM

16—MODERN WATER CLOSET

THERE IS ONE DAY ONE MAN APPEARS WITH WATER CLOSET.ITS PRESENTS THE SAME SMELL BAD THAN WATER NATURAL WATER CLOSET IN THE NATURE.THE REASON WE HAVE TO THINK ABOUT A NEW WC WITH POWER TO KEEP THE REJECT WITHOUT NOBODY SEES OR SMELLS IT.THE NEW KIND OF WATER CLOSET WILL HAVE 2 PARTS.THAT'S MEANT ONE BEFORE WHERE WE COULD SEE WATER,ONE AFTER WHERE THE REJECT WILL BE PLACED RIGHT AWAY IT WILL FALL TO BE PLACED RIGHT AWAY ON THE SECOND PART TO BE FLUSHED.WE 'LL AVOID TO FEEL A BAD SMELLING AND KEEP THE CLOSET ROOM CLEAN AND FRESH.

fortunegarcia@yahoo.fr

GARCIA FORTUNE'S TECHNOLOGY AND SYSTEM

17—SUN ENERGY

THE SUN ENERGY IS THE ONLY NATURAL ENERGY THAT THE WHOLE UNIVERSE IS USING SINCE THE BEGINNING OF THE UNVERSE USED WITHOUT ANY NEEDS OF REFULL. THE SUN ENERGY WITH IT CAPACITY TO REPRODUCT THE SAME TIME AND FOR THE SAME NEEDS EVERLASTING IS THE KIND OF ENERGY WE NEED TO STOP THE GABBAGE PRODUCTION OF THINGS.KEEPING ENERGY TO GIVE .THE SUN IS ALWAYS LIKE A STAR,THEY SAY SOMETIMES IT'S THE PLANET OF THE FIRE,A PLANET IN HIS WAY ,A KIND OF PLANET DIFFERENT FROM ALL PLANETS ,STRANGE,THERE IS NONE LIKE ITSELF ALL AROUND THE UNIVERSE ITSELF,IT IS THE ONLY ONE STAR COMING WHAT WE CALL IN OUR SYSTEM SOLAR,THE SUN.THE SUN BECOMES TO BE A PROBLEM FOR HUMAN PEOPLE. THE OZONE PART OF BLUE PLANET,THE EARTH IS DESTROYING AND THE X-RAY OR VIOLET RAY OF THE SUN PRODUCT GIVES A BURN TO OUR SKIN AND DEVELOP CANCER .THE LITTLE PART OF OZONE STAY IN OUR BLUE PLANET IS JUST TO KEEP US ALIVE WHEN THE EVERYTHING AND EVERYDAY SUN FLASH ON US ,WE FEEL LIKE WE ARE BURNT BY SOMETHING WE CAN'T CONTROL OR SOMETHING WE CAN'T STOP IF WARM EFFECT WILL GROW UP WITHOUT CONTROL OF HUMAN PEOPLE OR IF THE EARTH GO ON ITS WARMNESS FASTLY. THE SUN AND THE NATUNE GIVES US EXAMPLE THAT WE DON'T FOLLOW .EVERYTHING ,HUMAN PEOPLE CREATE BECOMES AND TURNS AGAINS OURSELVES HIS OWN DESTRUCTION LIKE CARS,BOATS,PLANES WITH THE BEST AND WRONG PART EVERYTHING AND ANYTHING WE TRY

TO CREATE PERFECT BECOMES UNPERFECT .NO WAY,BUT WE NEED NEW TECHNOLOGY AND WE ARE IN NEEDS,WE HAVE TO COMPLETE,FILL,FIND OR CREATE OUR NEEDS SOMETIMES,WE BELIEVE THAT ALL TIME WE CREATE COULD GO SO FAR AND CAN BE A GREAT HELP IN THE FIRST FOR US.NOTHING ELSE THAQT OUR NEEDS OF HAVING NEW CREATIONS UNPERFECT AND UNCOMPLETE.NOBODY BUT NO WAY IF WE CRY TO MAKE UP ON'S MIND WITH OUR TWO 2 HANDS BEFORE A CREATION ,TODAY WE WILL HAVE A PLANET GIVES US EVERYTHING AND EVERYTHING WE TAKE INSIDE OF IT RETURNS TO THE PLANET WITHOUT NO TROUBLE AND NO BAD EFFECT ,EVERYTHING IS CREATED ON EARTH TO HELP HUMAN PEOPLE TO LIVE SAFELY ,WE TURN IT IN SELF DESTRUCTION AGAINST THE PLANET WHICH TODAY DON'T NEED HUMAN PEOPLE NO MORE TO BE REGENERATES FASTLY AND POSSESS A NEW KIND OF CREATURES ON IT SURFACE LIKE THE DINOSAUR TIME,THE PRE HISTORICAL MEN TIME , THE HUMAN TIME AND SO ON,YES ,AFTER US WE REALLY DON'T KNOW WHAT THE PLANET EARTH WILL HAVE ON IT OR IT IS POSSIBLE THAT THE PLANET EARTH DECIDE ITSELF TO SAY ALONE FOR LONG MILLON OR MILLIARD YEARS TO BE AUTO SELF REGENERATED ,BUT THE UNIVERS WON'T FORGET US WITH ALL WE'VE DONE ON EARTH TRAVELLED ON THE THE UNIVERS BY THE WEB POWER THAT HUMAN PEOPLE CREATED.BEFORE US ,THERE IS LIFE AND AFTER US,HUMAN PEOPLE ,THERE WILL ALWAYS HAVE LIFE ON EARTH.

18—ROBOTS LIFE

SO MANY WORKS DONE BY HUMAN PEOPLE ARE COMPLETELY UNBELIEVALE.OUR IMAGINATION OF CREATION NEVER STOP.IN THE FUTUR NOW,WE PLAN TO BE HELPED BY NEW MECANICS ROBOTS COMPLETELY INDEPENDENT FROM US. SOMETIMES ,WE SAID PROGRAMMED THE WORK ,IT WILL BE DONE.NOW,DON'T NEED TO PROGRAM ANYTHING ,JUST TALK OR GIVE WORK PLANPAPER TO THE ROBOT , HE WILL READ IT AND ACCOMPLISH THE WORK AS YOU'VE DONE IT OR WROTE ON THE PAPER .DON'T NEED TO BE WORRY,THE WORK WILL BE DONE PERFECTLY LIKE IT WAS YOU DOING IT .

IN 2005,SO MANY GAMES REPRESENT THINGS AROUND THE NATURE IS CREATED IN ROBOT CARACTERITICS .THAT PROUVE OUR FUTUR WILL BE A COMPLETE FULL TIME HELPED BY ROBOTS.IN SO MANY INDUSTRIES WILL PRODUCE MORE AND THE MOST THEY WOULD LIKE TO PRODUCE WILL BE THEIR OWN WILL NOW,MECANICAL ROBOTS HANDS ALREADY EXISTED IN CERTAIN HIGH INDUSTRIES OF CARS MANUFACTURED ARE GREATLY USED IN PRECISION CREATION OF CARS,BOATS,PLANES.THAT KIND OF TECHNOLOGY PROVE WE COULD GO FASTER THAN THAT WE ARE NOW,THAT KIND OF HELP PERMIT US TO KEEP AWAY WORKERS FROM HIGH DANGEROUS DANGER OF OUR HIGHEST CONSTRUCTIONS PART OF THE INDUSTRIES TO COMBINE DIFFERENT PARTS.

TIME TO COMBINE THEM BECOMES SO STRENGHEN STRENGHLY FAST AND HAVE MORE PER HOUR IT'S THE

GAME OF INDUSTRIES PLAYING NOW AND TODAY AND IT WILL BW FOREVER.THESE ROBOTS PARTS COMPLETELY CONTROLLED BY PROGRAMMATION PROGRAMMED INDUSTRIALS PLAN COMPUTERS TO AVOID RISK OF BAD COMBINATION.ROBOTS MAKE ALREADY PARTS OF OUR LIFE NOW IN THIS BEGINNNING OF 21 ST CENTURY WHERE IN SOME HOUSE,WE CAN SEE SERVICES ROBOTS OR WAITRESSES ROBOTS,RECEIVERS ROBOTS,DRIVERS ROBOTS,HOUSE KEEPERS ROBOTS,THERE ARE SO MANY ANOTHER WORKS WITH PRECISION WORKS THAT OUR ROBOTS COULD REALIZE ,ROBOTS ARE STRENGHENLY PRESICE AND RESPONSIBLES IN OUR LIFETIME.

STEP BY STEP,THE CREATION OF OUR CREATION OF ROBOTS WILL BE MADE MORE AND MORE ADVANCED TO BE LIKE EXACTLY UNTIL LIKE HUMAN PEOPLE USED TO BE WHEN THEY RESPONSIBLE.SO STRANGER IT APPEARS,WE DEVELOP NEW TECHNOLOGY TO HELP US TOMORROW AND SAVE MORE AND MORE TIME TO TAKE CARE OF MORE THINGS MORE IMPORTANT IN OUR LIFETIME.WE STAND ABOUT A NEW LIFE STYLE STYLE COMPLETELY DIFFERENT FROM THE BEGINNING OF THE WORLD .SO MANY PEOPLE MAKE UP ON'S MIND IN HOPES TO REMIND THEIRSELF WE CAN PLAN A NEW STYLE OF LIFESTYLE DIFFERENT OF TODAY.WE WALK ON IT AND WE UNTIL ARRIVED TO GO THE GREAT CREATION OF THE CENTURY.ROBOTS ARE WORKING PART ALREADY OF OUR LIFESTYLE .THE PLAN TO HAVE THEM ANYWHERE,ANYPLACE COULD TAKE US BUSY,THERE ARE SO MANY WORKS VERY DIFFICULT FOR HUMAN PEOPLE AND HAVE ANBAD EYES ON THEM WILL BE HIGHLY DIFFICULT FOR US,BUT WE DRIVE ROBOTS AND LEAD THEM ON JUST EXACTLY THE PLAN WE REALLY WANT TO REALIZE ,IT WILL BE LIKE IT ALWAYS BE DONE IN THE OR OUR PAST WELL DONE DRIVEN BY US HUMAN PEOPLE,ROBOTS COULD ACCOMPLISH THEM WITH NO PROBLEM.ROBOTS LIFE ARE A NEW HOPE FOR HUMAN PEOPLE IN ALL THE WAY IN THE FUTURE.

fortunegarcia@yahoo.fr

GARCIA FORTUNE'S TECHNOLOGY AND SYSTEM

19—NATURAL WORLD ENERGY

TODAY,ALL GREAT POWER OF THE WORLD TRY TO BUILD AND DISCOVER NEW KIND OF ENERGY TO USE IN EARTH. TILL TODAY,THEY REALLY DON'T UNDERSTAND THAT THE TRUE ENERGY COMES FROM THE UNIVERS.TO ERADICATE WRONG ENERGY THAT MAKES THE WORLD SUFFERS,WE HAVE NOW COMPLETELY BAD ENERGY USING IN THE WORLD.THE NATURE IS DESTROYED BY THE WORLD POWERFUL USING OIL FROM CARS,INDUSTRIES IN MAJORITY.THE OIL BURNT AND BUILT THE WARMNESS THAT WE KNOW DESTROYING THE EARTH PROTECTION.IN FRANCE NOW,WE TALK ABOUT A SYSTEM OF PLANT PLACED ALL AROUND THE CITY TO CAPT THE CARBON,YES,TOO MUCH CARBON OR CO_2 OR CARBONIC GAZ CREATING THE WARMNESS EFFECT INSIDE THE WORLD.THIS SYSTEM IN FRANCE COULD BE REGENERATED THE WORLD BY THOSE SPECIAL PLANTS CAPTING THE CO_2 AND TRANSFORM IT IN O_2,THE MOST GAZ THAT THE WORLD NEEDS,OXYGEN. IF ALL AROUND THE WORLD ALL GREAT COUNTRIES LIKE USA,JAPON,CHINA AND DEVELOPING COUNTRIES WILL DO THE SAME,YOU IMAGINE WHAT COULD ARRIVE TO THE WORLD,THE PLANET EARTH WILL SELF AUTOMATICALLY REGENERATED,IT WILL BE GREAT.IN USA,WE DISCOVER SO MANY KIND OF ENERGIES WE STUDY TO USE ONE OF THEM OR ALL TO TOGETHER TO HAVE A HIGH ANSWER ASSOCIATED TO THE PLANTS ABSORBING CO_2 WILL GIVE A POWERFUL,A STRANGE EFFECT TO THE WORLD TO LET US SPEND SOME MILLIARD OF YEARS INSODE OF IT.HUMAN PEOPLE MUST STRICLY UNDERSTAND THAT EACH TIME WE

GO ON WITH THE SAME DESTROYING OUR OZONE,THE PLANET EARTH IS TRANSFORMING ITSELF IN A BOMB TO KILL OURSELF OR DESAPPEAR US FROM THE WORLD.WE USE NATURAL ENERGY PROVIDES FROM SUGAR CANE,WATER AND SOME VEGETABLES IN BRASIL,CUBA IS TRYING TO DEVELOPED A SPECIAL ONE WITH SUGAR CANE,USA RESEARCHERS ARE DEVELOPING THE SUN ENERGY AND ALL AROUND THE WORLD,THE SUN ENERGY BEGINS TO BECOME THE FIRST ONE FOR EVERYRHING,THIS MAKES US UNDERSTOOD THAT THE SUN ENERGY COULD BECOME OR COULD REPLACE TOTALLY IN THE NEAR FUTUR ALL OUR INSTALLATION CREATING CO_2 AT GREAT LEVEL.

IMAGINE A WORLD WITH A COMPLETE ENERGY,PURE NEW BRAND ENERGY,THAT WILL HELP THE WORLD TO BE SAFE AND QUIET TO BE REGENERATED DURING SOME TIMES TO RUN ON ITS FIRST BEARTH LIKE A BABY BORNS.

FRENCH,CHINESE MAY BE JAPONESE WORK ON THE SAME ENERGY TO USE IT TOGETHER.THERE ARE SO MANY ANOTHER COUNTRIES TRYING TO DO THE SAME TO SAVE THE WORLD BLUE PLANET.WE CONSISTENTLY WORK ABOUT WATER,WIND,THE SUN EFFECT EVELASTING IS COMPLETELY DEVELOPED HIGHLY BY ALL GREAT NATIONS ALL AROUND THE WORLD,IF OUR WORK IS UNITED,THE EFFECT WILL BE ONE,BUT IF OUT WORK ISN'T UNITED,IT WILL BE LIKE A MINI SUGAR SPOON IN GREAT GLASS OF WATER.SO MANY COUNTRIES TRY TO USE OIL FRO ANOTHER ELEMENTS,NATURAL ELEMENTS FROM EARTH FROM SO MANY THINGS WE EAT OR REJECT.

ONE IDEA COME THROUGH MY MIND RIGHT NOW,LISTEN UP,THE WORLD PRODUCE ONE MILLION AND OVER OF TONES OF GABBAGES BY HOUR,IF WE COULD TRANFORM THE MILLION TONES OF GABBAGES THAT THE WORLD USED EVERYDAY IN A NATURAL OIL REJECTING THE SAME O_2 WHEN IT IS USED IN OUR LIVING SPACE.LIKE GABBAGE,WE COULD FIND BY GREAT STAIR ALL OVER THE WORLD.

CONTROL THE FUTUR OF THE FIRST ENERGY THAT WE USED,COULD BE DIFFICULT FOR US IF WE DON'T STOP TO USE THEM RIGHT NOW.

THE WARMNESS OF THE WORLD WILL PRODUCT THE DEFREEZENESS OF THE NORTH POLE OR ANTARTIC COULD BE DANGEROUS,SO HIGHLY DANGEROUS FOR US THAT THIS WATER COULD EMERGE ALL THE WORLD LETTING ONLY SOME COUNTRIES.REMIND YOURSELF,THE NORTH POLE REPRESENTS THE GREATEST WATER RESERVE OF THE WORLD,THE ONLY ONE THAT WE HAVE,WATER RESERVE FOR THE WORLD.THE FAST DEFREZENESS OF NORTH POLE WILL GIVE MORE WATER FASTLY THAN YOU EVER KNEW AND SAW IT IF THE NATURAL POWER DEFREEZES IT SLOWLY LIKE IT IS AND WHEN THE NATURAL POWER UNDERSTAND THE WORLD NEEDS IT STEP BY STEP.THAT FAST DEFREEZENESS COULD A TROUBLE FOR THE WORLD WATER CAPACITY AND THAT SO MUCH WATER COULD PRODUCT BAD EFFECT FOR THE WORLD.NEW ,WE PROVIDE FROM THE NATURE IS A SOLUTION HUMAN PEOPLE HAVE TO USE SINCE THE FIRST STEP OF THE WORLD AFFAIR.

THEY ALWAYS SAID,IT'S NEVER THROUGH LOVE WE ARE GOING TO CORRECT A MISTAKE WE'VE DONE SOMEWHERE,THE WORLD IN THIS CASE COULD SHOW AND PROVE US THE CONTRARY,OUR LOVE FOR THE WORLD,OUR DEAR PLANET EARTH COULD BE REGENERATES AS FASTLY AS POSSIBLE,SO FASTLY THAT US HUMAN PEOPLE WILL DOUBLE OUR ABILITY TO LIVE MORE MILLIARD OF YEARS ON THIS PLANET,OUR DEAR GREAT HOUSE,THE ONLY ONE INSIDE OUR UNIVERS CREATED SPECIALLY FOR US HUMAN PEOPLE,THE PLANET IS WORTH OUR LOVE AND TO BE SAVED BY US HUMAN PEOPLE BY CREATING A SYSTEM OF SALVATION,A FAST SYSTEM OF SALVATION TO DRIVE THE WORLD,THE PLANET EARTH TO ITS FIRST STEP,THE UNIVERS IN THIS CHANGEMENT WILL ACCOMPANISH US AND WILL SEND MORE POWER TO HELP THE WORLD TO LIVE AND ACCEPT TO KEEP US DURING SOME MILLIARD OF YEARS AGAIN.

A MISTAKE, WE BEGIN TO PAY BY THE NATURAL EFFECT. WHEN IT IS HOT,IT SO HOT.COLD,IT IS TOO COLD.WHEN THE SEA IS BECOME DANGEROUS,IT KILL SO MANY PEOPLE LIKE THE TSUNAMI DONE IT IN THE BEGINNING OF THE 21RST CENTURY,KILLING AN AROUND 200 THOUSAND PEOPLE (2 HUNDRED THOUSAND),WE COMPLAIN IT DURING SO MANY DAYS,SO MANY COUNTRIES REFUSE TO SEE,THEY HAVE TO MINIMIZE THEIR PRODUCTION WHICH PUT HUNGRY THE WORLD.THE BAD GAZ PRODUCTION OF THE INDUSTRIES WORLWIDE GROW UP THE ABILITY OF NATURE TO BE WORST EFFECT IN THE WORLD.TO MAKE MONEY ALL GREAT NATIONS REFUSED TO SEE THAT THE BLACK GOLD PETROLEUM IS ENDED FOR THE WORLD NOW AND MUST BE CATEGORICLY REPLACED,YES CATEGORICLY REPLACED BY A NATURAL OIL GIVING O2 TO O2 TO MAXIMIZE THE POWER OF LIFE OF THE WORLD.

MONEY,MONEY,MONEY,MONEY OF THE BLACK GOLD PETROLEUM MAKE OUR WORLDWIDE LEADERS REFUSED TO UNDERSTAND THAT DESTROYING OUR ONLY HOME,THE PLANET EARTH IS DYING STEP BY STEP AND UNDERSTAND AND THAT TRUE THE PLANET WILL NOT,BELIEVE IN WHAT YOU'RE READINDG TODAY,THE PLANET WILL NOT ABSOLUTELY REFUSED TO SACRIFICE ITSELF BY ITS DESTROYING WITH US TO DESAPPEAR FOREVER IN THE UNIVERS,WE COULDN'T UNDERSTAND,WE WILL BE DESAPPEARED FOREVER TO LET OUR PLACE TO NEW KIND OF LIFE ON EARTH,REMIND YOURSELF,LIFE ALWAYS EXISTED ON EARTH REPLACING STEP BY STEP DIFFERENT KIND OF BEING ON IT,AFTER OURSELF,THERE WILL BE ANOTHER KIND OF LIFE,WE DO MUST UNDERSTAND THAT.

LIFE SPENT IN THE SECOND RANGE PLACE.IT'S LIKE SOMEONE BUILD AN HOUSE TO LIVE ,TWO (2) DAYS AFTER,HE BEGINS TO DESTROY HIS HOUSE AND 10 YEARS AFTER,THE HOUSE FALLS DOWN ON HIM HE FINISHES TO REALIZE,HE HAS NO WAY ELSE TO GO,IT IS WHAT WE ARE DOING NOW INTHE WORLD.

THE SAME THING FOR THE WORLD ,ONE DAY ,WE WILL REALIZE THAT WE ARE DESTROYING OUR ONLY GREAT UNIVERSAL EARTH HOME AND IT WILL BE MAY TO LATE IF DON'T UNDERSTAND THAT NOW

OUR WORLDWIDE LEADERS MAKE UP ONE'S MIND ABOUT THE PETROLEUM BEGINNING TO BECOME UNFOUND IN THE WORLD,THEY HAVE LIKE IDEA TO LOOK FOR IT IN THE UNIVERS,THE PRINCIPAL SOURCE OF PETROLEUM .THAT MEANT THEY DECIDE THEMSELF TO DESTROY THE WORLD COMPLETELY COMPLETE.

THE PERFECTION IS IN ALL,TO BE PERFECT IS SOMETHING THAT US ,HUMAN PEOPLE MISSED AT THE BEGINNING OF OUR FIRST STEP ON EARTH AND THE UNIVERS HAVE TO DRIVE US,HUMAN PEOPLE IN THAT PERFECTION

THERE IS NO HOME LIKE THE WORLD,THE BLUE PLANET ALL AROUND THE UNIVERS,IT IS THE ONLY ONE WITH O2 IN WHAT WE COULD LIVE IN WITH NO PROBLEM.REMIND YOURSELF TO END THAT THERE IS NO BLUE PLANET,THE PLANET EARTH LIKE OURS EXITED ALL OVER THE UNIVERS.

fortunegarcia@yahoo.fr

GARCIA FORTUNE'S TECHNOLOGY AND SYSTEM

20—REGENERATES CELLULARS

SO MANY SICKNESS DEPEND IN AN CHIRURGICAL INTERVENTION TO CARE IT.MEDECINE DAY AFTER DAY BECOMES SO STRONG AND MORE DIRECT,DIRECT CARE TO SOLVE BODY TROUBLES.IN THE FUTURE,WE ARE GOING TO ASSIST TO A NEW KIND OF MEDECINE MORE IMPORTANT THAN THAT WE HAVE TODAY TO COME TO A FAST RESOLVATION OF THE BODY.SOMEONE HAS A CUT SOMEWHERE IN HIS BODY.SOMETIMES,WE HAVE TO COME TO A FAST CHIRURGICAL INTEVENTION WITH A NEW KIND OF CELLULARS ABLE TO RESOLVE ALL CHIRURGICAL TROUBLES AUTOMATICALLY FOUND OR INJECTED IN YOUR BODY.INJECTED,THOSE REGENERATES CELLULARS COULD BRING A CHIRURGICAL INTERVENTION DIRECTLY IN THE HUMAN PEOPLE,IN THE HUMAN BODY AT THE EXACT PART.THESE NEW KINDS OF CELLULARS NEVER HAVE BEEN NEVER EXISTED YET OR FROM TODAY,BUT HAVE TO BE CREATED LIKE PROCESSOR INSIDE A COMPUTER OR EVERYTHING EXISTED LIKE THAT.

TOMORROW,WE WILL HAVE PROCESSORS IN EVERYTHING THAT YOU'RE USING,IN REALLY EVERYTHING,EVERYTHING THAT HUMAN PEOPLE WILL USE FOR HIS BEST LIFETIME.

THE BEST CARE FOR HUMAN PEOPLE FOR THE FUTURE.

VERY IMPORTANT :

IF YOU ARE INTERESTED OF MEDICAL ADVANCED TECHNOLOGY,YOU COULD CONTACT ME TO BUY IDEAS ON MEDICAL ADVANCED TECHNOLOGY NOT WRITTEN IN THIS BOOK ON

fortunegarcia@yahoo.fr

GARCIA FORTUNE'S TECHNOLOGY AND SYSTEM

21—FUTURE CARE

TO CARE PEOPLE IN THE FUTUR,A NEW KIND OF MEDICAMENT OR MEDICARE WILL BE AUTOMATICALLY IN USED.IMMEDIATELY THAT YOU FEEL YOURSELF UNWELL,YOU JUST HAVE A CONSULTATION IN SAYING WHAT YOU FEEL WRONG INSIDE OF YOU.THIS KIND OF MACHINE IN PUTTING YOUR OR ONE OF YOUR FINGER COULD MAKE YOU A FAST BLOOD TEST AND REACT IMMEDIATELY BY GIVING YOU MEDECINE FASTLY IF NECESSARY OR SEND YOU TO BUY IT IN A DRUGSTORE. THIS DOCTOR MACHINE IS SPECIFIC AND DIRECTLY HIGH LEVEL WITH PRECISION IN EACH POINT OF VIEW IT WILL PRECISE YOU TO DO FOR YOUR CARE.BY ONLY CELLULARS THROUGH X-RAY OR SIMPLE LAZER RAY,PEOPLE COULD RECEIVE CARES AND BE SAVED FOR AND AGAINST THE SICKNESS OR TROUBLESBODY THEY HAVE.SOME NEW WAYS TO CARE PEOPLE IS LOOKING OR BEING SEARCHING TO HELP THE FAST CARE AND EASY SOFT CARE.

LIFETIME WILL BE EVERLASTING BY A NEW KIND,A NEW CELLULAR KIND KEEPS HUMAN PEOPLE YOUNG FOREVER.

SCIENTISTS FOR,IN THIS CENTURY ARE READY TO PREPAIR A GOOD HEALTH CARE FOR THE WELL BEING OF EVERY HUMAN PEOPLE.

fortunegarcia@yahoo.fr

GARCIA FORTUNE'S TECHNOLOGY AND SYSTEM

22—GLOBO CAR

A NEW CAR WITH 3 HIGH ENERGIES CAPACITIES TO RUN AND PROTECT,SAVE THE WORLD.THAT CAR WILL BE LIKE ALL CAR WITH THE SAME BEAUTIFUL DESIGN,MORE ADVANCED,THE CAR WILL HAVE:

THE SUN ENERGY FIRSTLY INSTALL IN THE HEAD OR IN ALL THE CAR TO CAP MORE ENERGY CAPACITY DURING THE JOURNEY.

SECONDLY,THE CAR BY ITS RUNNING PRODUCT ENERGIES WAS ABLE TO CAPT IN A BOX TO HELP WHEN THERE IT'S DARK.

AT LAST,A NEW KIND OF BATTERIES WILL BE DEVELOPPED FOR EVERLASTING TO HELP THE CAR MOVE DAY AND NIGHT WITH NO PROBLEM.

THE WATER ENERGY COULD BE USED TOO

THE PROPAN ENERGY IS ALREADY IN USED LIKE IN CUBA,MEXICO CAN BE WITH ONE BETWEEN THE 3 ENERGIES TO HELP THE CAR MOVING FAST.

GOOD ENERGIES

THE GOOD ENERGIES THE WORLD COULD USE ARE

—MOTRICITY ENERGY

—PROPAN ENERGY

—WATER ENERGY

—BATTERY EVERLASTING ENERGY

—SOLAR ENERGY

COMBINATED TOGETHER IN ONE NEW CREATION, THE CAPACITY FOR EXAMPLE A CAR CAN BE GREATLY SUPERIOR BY THE PRESENT EVERLASTING ENERGY PEOPLE USE TO COMBINATE AND DESTROY THE WORLD. NOW, PEOPLE HAVE POSSIBILITY TO HAVE AND PROTECT THE WORLD BY GOOD ENERGIES IN EVERYTHING WILL BE WELL FUNCTIONNED WHEN EVERYTHING WILL NEED ENERGY TO WORK. THE CAPACITY OF THESE ENERGIES ARE GREATS AND CAN BE FOREVER WHERE PEOPLE WILL ECONOMIZE MONEY TO DO SOMETHING ELSE IN THE WAY TO FORGET OIL AND GAZ SYSTEM.

IMAGINE SOMETHING WORKING WITH NEW ENERGIES AND WORK FOREVER AT THE SAME TIME, THAT WILL BE, THAT WILL HELP THE WORLD AND YOU TO SAVE SOME MONEY AND FOR YOUR HEALTH, A PROTECTION LIFESTYLE.

GARCIA FORTUNE'S TECHNOLOGY AND SYSTEM

fortunegarcia@yahoo.fr

23—UNPLACED PERSONAL COMPUTER

A NEW KIND OF COMPUTER KEEPING ANY PLACE ANYWHERE AND CAN BE PLACED AND GO AWAY WITH HIS POSSESSER TO BE USED ANYWAY AT ANYTIME .WHAT KIND OF COMPUTER IT WILL BE ,WHAT FOR ,TO DO WHAT.

ANYBODY,YES,ANYBODY AT THE CONCEPTION OF THIS MATERIAL COULD USE IT BY VOICE COMMAND,FOR BLIND PEOPLE WHO COULD TALK,A DIRECT POINT WILL BE PLACED IN ONE OF HIS FINGERS TO BE CONNECTED TO THE TOTAL ABILITY OF THE BRAIN ABLE AT THIS MOMENT TO BE CONTROLLED BY MIND DIRECTLY WITHOUT TALKING IF THE PERSON WANTING.

ANYBODY CAN'T LIVE WITHOUT IT AND WAIT FOR IT TO BE GROWTH AND USED IN CONSTRUCTION OF GREAT LEVEL. ALL THE WORLD CAN HAVE IT AND BUY IT TO POSSESS IT BY MILLIARD OR IT,IT TOOK ANY PLACE AND COULD DEVELOPED ANYWHERE.BY MILLIARD OF IT FOR EACH PERSON ALL AROUND THE WORLD AT A BETTER PRICETHAT STRONGLY FOR EVERYONE AND THE TROUBLE OF UNCAPATCITY OF SOME COUNTRIES TO BE DEVELOPED AND CERTAINLY THE PROBLEM OF POOR UNIVERSITY YOUNGS PEOPLE WILL BE RESOLVED,TO DEVELOP NEW TECHNOLOGY OF COMPUTERS AND INTERNET OR WEB WILL BE RESOLVED IN GREAT WAY.EVERYONE COULD USE IT WITH NO PROBLEM HOME,OUTSIDE IN THE OFFICE,IN

HOLIDAYS,ANYWHERE SEE IT IS NOT HEAVY AND IT IS JUST HANDLE IN YOUR HAND LIKE A JEWEL AND IT ABILITY IS STRONG AND WITHOUT DANGER WHEN OR DURING YOU BRING IT.IT IS ABLE TO DO SO MANY THINGS IN THE SAME TIME LIKE YOUR OWN DIRECT PERSONAL COMPUTER SEAT ON THE DESK OR TABLE AT HOME,AND GREAT CAPACITY LIKE GROWTH GREATLY THE SCREEN AS LARGELY AS POSSIBLE TO DO WHATEVER YOU WANT ANYWHERE THAT IS STRONGLY THE CAPACITY OF THE USER COMPLETELY. THE MORE IMPORTANT IT WILL BE THAT UPC WILL ASK ANYTHING.JUST A LITTLE BOX WILL BE ACTIVED TO CONTROL THE GROWTHFUL OF THE SCREEN.THE RESON WHY ANYBODY,ANYWHERE,ANY COUNTRIES COULD USE IT WITH IT SIMPLE WAY SO EASILY,YES,WILL BE ABLE TO BUY IT AT A LOW PRICE TO HELP THEIR PEOPLE COMMNICATE BETWEEN THEMSELF ALL THE WORLD.

THAT MAIN PORTABLE LITTLE BOX WILL ACTIVATE THE SCREEN,THE TOUCH WRITER OR TYPE WRITER AND THE BRAIN BOX WILL BE SPECIALLY CONSTRUCTED WITH CAPACITY FOR LONGTIME LIFETIME AND CAN BE MODIFIED STEP BY STEP THAT TECHNOLOGY WILL GROWTH ALL AROUND THE WORLD.

THE SPEAKER,THE PRINT WILL BE CREATED A WAY SPECIALLY TO BE ADAPTED TO THAT NEW CREATION.

THE MOUSE WILL BE ACTIVATED TOO BY THE MOUSE .COMPUTER OR EVERYTHING CAN BE ASKING TO THE UPC BY TOUCHING ONLY BY DRAWING THE FORM YOU WANT ON IT LIKE IT EXISTS EARLY.

THAT MOUSE PERSONAL COMPUTER WILL ACTIVATE THE LAZER COMPUTER DRAWING BY ITSELF WITH GREAT AND HIGH TOUCH PRINT.EVERYTHING WILL BE IN COLOR DIFFERENT DESIGN FROM THE LAZER WRITING.

THE UNPLACED COMPUTER SETTING WON'T HAVE PLACE,IT JUST WILL BE SUPPORTED BY SOMETHING

LIKE A WATCH IN YOUR HAND WILL BE THE BRAIN,THE BOARD,THE MOUSE,ALL THOSE THINGS DRAWING BY A LAZER DRAWING THE SCREEN TOTALLY WHERE IT IS READY TO BE USED FOR ALL SET LIKE TV,CELL PHONE WITH GPS INSIDE THE SUPPORTED WILL HELP YOU TO REALIZE A LOT OF THINGS BY VOICE COMMAND OR DIRECTLY BY YOUR HANDS. THAT UNPLACED PERSONAL COMPUTER WILL BE COMPLETED WITH ALL NEW TECHNOLOGY AND SYSTEM ABILITY TO BE STRONGLY ABLE TO SERVE TO DO EVERYTHING THAT AN HUMAN PEOPLE COULD DESIRE AT HOME OR AT HIS OFFICE.

THIS IS THE HIGHEST AND THE MAIN TECHNOLOGY FOR A FAST WORK INSIDE IN OUTSIDE THE HOUSE CONTROLLED TOTALLY JUST IN OUR HAND TO GO EVERYWHERE.

fortunegarcia@yahoo.fr

GARCIA FORTUNE'S TECHNOLOGY AND SYSTEM

24—DIRECT SCANER FOR PERSONAL COMPUTER

TILL THE END OF THE LAST CENTURY,WE WORK FASTLY,MORE FASTLY AND WE NEED MORE POWERFUL TO GO STRONGLY TO NEW WORLD SO EASIER THAN ANY WORLD EXISTED AROUND THE UNIVERS.WE TALK ABOUT THE UNIVERS,WE TALK ABOUT IT THAT WE WANT IT AND WE WORK TO GO TO IT,THAT MEANS ONE DAY OUR WORLD WOULD COME SO SIMPLE THAT WE'LL HAVE MORE TIME TO DO MORE THINGS THE TIME WE WANT WHATEVER WE THINK ABOUT IT.WHEN WE ARE THERE,IT'S THE WORLD POWERFUL SHOWING .

LET'S TALK ABOUT THE DIRECT SCANER COMPUTER.IT'S A MINI SCANER WHICH WILL BE CONNECTED TO THE COMPUTER,INSIDE THE COMPUTER OR THE LAPTOP TO COPY EXACTLY OR IN ANOTHER WRITING THE SAME TIME TEXT SOMEONE WRITES ON A SHEET.LIKE THE PHOTOCOPY MACHINE BUT MORE LITTLE.THAT'LL HELP SOMEONE TO GO FASTLY,FASTER TO RECEIVE A TEXT AND AFTER TO CORRECT IT ONLY.

THAT DIRECT SCANER FOR PERSONAL COMPUTER IS THE MACHINE COPY WILL BE SPENT ON THE COMPUTER DIRECTLY BY THE COMPUTER TO GO ON THE WORK BY SOMEONE OR PUBLISHING HOUSE .

fortunegarcia@yahoo.fr

LAZER USING

IN THIS BOOK, I'VE TALKED ABOUT THE ABILITY OF LAZER RAYS IN ALL KINDS OF PART OF LIFETIME FOR EXAMPLE AT HOME,FOR OUR SECURITY,FOR OUR SELF MEDECINE,DOCTOR CARE,LIKE OUR OWN NEEDS IN EVERYTHING WE'RE DOING IN EACH PART IN OUR LIFETIME.THERE IS NO TROUBLE ABOUT TO KNOW THERE WILL BE DIFFERENT KIND OF LAZER FOR DIFFERENT KIND OF WORKS WE ARE GOING TO REALIZE IN OUR LIVING TIME.A SURE WARNING WILL BE PLACED ON EACH MARERIALS USING LAZER TO SAY IF IT SOFT OR HARD LAZER ABLE TO CUT OR KILL.

LAZER WILL MAKE PART OF OUR LIFETIME SO CURRENTLY FOR A EASY LIFE SET.

fortunegarcia@yahoo.fr

GARCIA FORTUNE'S TECHNOLOGY AND SYSTEM

A SPECIAL WORD TO END

GARCIA FORTUNE'S TECHNOLOGY AND SYSTEM

WHEN I BEGIN THIS BOOK LIKE HANDWRITING SEVERAL YEARS AGO,I REALLY DIDN'T KNOW HOW TO PUBLISH IT AND HOW I AM GOING TO DO TO PUBLISH IT,ONLY,THE ONLY THING I KNOW,I KNEW THAT I WANT TO PUBLISH THIS BOOK FASTLY IN STRANGER COUNTRIES LIKE USA,CANADA OR FRANCE FOR FRENCH BOOK.I REALLY DIDN'T KNOW WHAT TO DO IN THE AVENIR OF MY COUNTRY WAS VERY UNCERTAIN FOR PEOPLE LIVING INSIDE OF IT WHERE CITIZEN REFUSED TO PUT HANDS TOGETHER AND WORK FOR THE SAFE AND WARM FOR THE COUNTRY THAT ONLY CITIZEN OF FOREIGN COUNTRIES ONLY KNOW AND KNEW THAT WHAT WORTH THIS COUNTRY THAT A LOT PEOPLE OUTSIDE OF THE COUNTRY WOULD LIKE TO LIVE IN. TO BE BACK IN THE PUBLISHING OF "100% NEW TECHNOLOGIES AND SYSTEMS" GARCIA FORTUNE'S TECHNOLOGY AND SYSTEM,I DECIDED MYSELF TO SEND VIA THE WEB ALL TITLES WITHOUT ANY DESCRIPTION OF ANY OF THEM WITH THE HOPE THAT NOBODY COULD REPRODUCE THE CREATIONS,I'VE DONE IT AND I SAW THAT A LITTLE PART OF THEM WERE USED.AFTER TWO PUBLISHING BOOKS OR NOVELS WITH XLIBRIS WHERE I INCLUDED SOME NEW TECHNOLOGIES AND SYSTEMS INSIDE THEM,

"HER'S FLYING CAR" AND

"THE LAST EAGLE"

YOU CAN PURCHASE THEM ON

WWW.AMAZON.COM

I SAID MYSELF IT IS TIME TO PUBLISH "100% NEW TECHNOLOGIES AND SYSTEMS" A COMPLETE DESCRIPTIVE TECHNOLOGIC NOVEL OF EACH TITLES I'VE SENT ON THE WEB

NOW THE REAL TIME WILL BE TRUELY ARRIVES ON THE WEB FIRST AND AFTER THE PAPER BACK WILL COME SOONLY AND WITH NO MERCY OF THE BEST OF BEST COULD BUY IT ON THE WEB FIRST,IT WILL BE AVAILABLE AT THE ONLY PRICE FOR EACH " 100% NEW TECHNOLOGIES AND SYSTEMS" THAT PEOPLE ALL AROUND WILL BUY FOR : $ 99,99 US

YOU REALLY WANT TO KNOW WHAT THE FUTURE WILL BRING OR RESERVE TO US ,THIS IS THE BEST BOOK TO READ

EACH TITLES OF THIS BOOKS IS FOR SELLING ,YOU ARE LOOKING FOR NEW CREATION TO PRODUCE AND YOU WOULD LIKE TO INVEST IN HIGH INDUSTRY PRODUCTION

JUST CONTACT ME ON:

fortunegarcia@yahoo.fr

ALL INSIDE THIS BOOK WAS WROTE BY GARCIA FORTUNE

CHINA

NOW,IN THE WORLD,CHINA TECHNOLOGICALLY IS THE COUNTRY DOING THE MOST OF RESEARCHES IN THE WORLD.THEY DEVELOPPED ALL DIFFERENT KIND OF TECHNOLOGIES THE MORE AND MORE DEVELOPED EACH

ANOTHER.TODAY,IT IS VERY IMPORTANT FOR ME,THE WRITER OF THIS BOOK TO SALUATE THEIR WORK AND WHAT THEY ARE GOING TO BRING TO THE WORLD THE MOST ADVANCED IN THIS CENTURY WILL SURPRISE A LOT OF GREAT POWERS ALL OVER THE WORLD.

THE FUTUR WILL BE OURS,JUST LIVE IT

THIS BOOK IS WRITTEN BY THE AUTHOR IN THE GREATEST SECRET POSSIBLE JUST TO HAVE AN ORIGINAL WORK.

EACH DAY OF MY LIFE,I THOUGTH ABOUT WHAT TO WRITE INSIDE THIS BOOK.I LIVED,WORKED,ATE,LOOKED AT TV WITH MY LAPTOP IN FRONT OF ME JUST TO WRITE YOU EACH IDEAS COMING,PASSING IN MY MIND.

I DO BELIEVE IN THIS ORIGINAL WORK AND I DO BELIEVE TOO THAT THAT WORK WILL SERVE TO A GREAT CHANGEMENT OF THE WAY WE,HUMAN PEOPLE WILL LIVE IN THE WORLD.

VERY IMPORTANT

GARCIA FORTUNE'S TECHNOLOGIC WORK

fortunegarcia@yahoo.fr

WHAT IS SURPRISING ME THE MORE IS THE ABILITY OF TECHNOLOGIC WORK,IDEA NOT YET FINISHED,ENTERING IN CONTACT WITH IT,MY MIND COULD THINK ABOUT THE COMPLETE FINAL TECHNOLOGIC WORK AS FASTLY AS POSSIBLE.TO CONTACT GARCIA FORTUNE

fortunegarcia@yahoo.fr

ALL MYSELF IS TECHNOLOGY

THE POWER OF CREATION IS BETWEEN OUR HANDS.

EVERYTHING IMPOSSIBLE,GOD TOOK THEM,ALL THINGS POSSIBLE ARE STAYING FOR US HUMAN PEOPLE.

THIS IS A STEP THAT WE HAVE TO TAKE.

I KNOW IN PUBLISHING THIS BOOK,A LOT OF THINGS ARE GOING TO BE CHANGED IN THE WORLD,THE WAY,WE HUMAN PEOPLE THINK IS GOING TO HAVE A GREAT NEW BRAND POSITIVE MIND COULD MAKE YOU SEE THE THINGS,ALL AROUND YOU DIFFERENTLY THAN THE FIRST STEP YOU HAVE LIKE A BABY.NOW, HUMAN PEOPLE WILL HAVE A LOOK ON EVERYTHING THEY WILL USE AT HOME,NONE OF WHAT WILL BE,THE MORE RICH IMPORTANT THING IS THAT THE WORLD IS ABLE TO KEEP US AGAIN BY OUR NEW WAY OF LIFE.

GARCIA FORTUNE

WELCOME IN THE WORLD OF :

GARCIA FORTUNE'S TECHNOLOGY AND SYSTEM

fortunegarcia@yahoo.fr